Lecture Notes in Mathematics   1778

Editors:
J.-M. Morel, Cachan
F. Takens, Groningen
B. Teissier, Paris

T0240730

Springer
Berlin
Heidelberg
New York
Barcelona
Hong Kong
London
Milan
Paris
Tokyo

Hubert Kiechle

Theory of K-Loops

Springer

Author

Hubert Kiechle

SP Geometry and Discrete Mathematics
University of Hamburg
Bundesstrasse 55
20146 Hamburg, Germany
e-mail: kiechle@math.uni-hamburg.de

Cataloging-in-Publication Data applied for

Die Deutsche Bibliothek - CIP-Einheitsaufnahme

Kiechle, Hubert:
Theory of K-loops / Hubert Kiechle. - Berlin ; Heidelberg ; New York ;
Barcelona ; Hong Kong ; London ; Milan ; Paris ; Tokyo : Springer, 2002
 (Lecture notes in mathematics ; 1778)
 ISBN 3-540-43262-0

Mathematics Subject Classification (2000): 20N05

ISSN 0075-8434
ISBN 3-540-43262-0 Springer-Verlag Berlin Heidelberg New York

Springer-Verlag Berlin Heidelberg New York a member of BertelsmannSpringer
Science + Business Media GmbH

http://www.springer.de

Typesetting: Camera-ready TEX output by the author

SPIN: 10866597   41/3142/DU - 543210 - Printed on acid-free paper

To my beloved wife Maria

Preface

This book contains the first systematic exposition of the presently known theory of K-loops. Besides this, it presents some new results and many examples. Furthermore, the two most important applications are described in detail. Since about ten years the subject of K-loops has grown rapidly, so it seemed reasonable to put things in order.

There are not many books on quasigroup and loop theory. The oldest are BRUCK's [20] and BELOUSOV's [10]. PFLUGFELDER's [94] is used as a general reference for the collection of survey articles [24]. Some basic loop theory is also contained in books on projective planes, such as PICKERT's [95]. More specialized, but with very different orientation are SABININ's recent [107] and UNGAR's most recent [119] publications. Therefore, most of the material covered has not appeared in book form before.

Chapters 1–6 try to unfold the theory in a coherent and self contained way, and could be used as a text. The only prerequisite is basic algebra, in particular, group theory. With very few exceptions[1] complete proofs are given. Examples are later given in Chapters 9,11,12. However, for a course Chapter 12 can be developed as needed to enrich the theory with examples.

Chapters 7–11 are more like research notes, and only partially suitable for the classroom. Still, proofs are concise, but complete. Additional prerequisites for Chapter 9 are some linear algebra and matrix groups over ordered fields, a bit of ring theory for Chapter 11, and a basic knowledge of special relativity in Chapter 10.

While Chapters 2–7 build up almost linearly, later sections have only the following additional dependencies: Chapter 9 and Chapter 11 both use Chapter 8 for some special results. Chapter 10 builds on Chapter 9, but not on Chapters 7,8.

We have opted not to include exercises. However, some of the remarks can be used as such.

---

[1] These exceptions are some results on the isotopy of Bol and K-loops, which are not used elsewhere in the book.

The appendix is meant to put the material into a historic perspective, and should be seen as a supplement to the introduction. It is definitely not the result of conclusive research on the history of the subject.

Enumeration of theorems is straightforward. Occasionally equations are numbered (i), (ii), etc. These numbers will only be referred to locally, i.e., within a section. Remarks are not enumerated, for they are not cited in the text.

Finally, it gives me great pleasure to thank Andrea Blunk, Helmut Karzel, Wen-Fong Ke, Michael K. Kinyon, Sebastian Rudert and an anonymous referee for many valuable comments and suggestions. I am especially grateful to Alexander Kreuzer for carefully reading the manuscript. Without his help many typos, obscure statements, and inconsistencies would have gone undetected.

Kaufering, September 2001   Hubert Kiechle

The graph on the cover depicts the relationship of various structures discussed in the book.

Contents

Introduction

K-loops are non-associative generalizations of abelian groups. They made their first appearance early in the development of quasigroup and loop theory. Over the years, a few papers in this area were published, but the related structure of Bol loop has drawn much more attention. This changed when UNGAR [115, 116] showed that the set of admissible velocities with the addition of velocities in special relativity forms a K-loop. UNGAR's discovery sparked a rapid development of the theory of K-loops in the last decade.

Besides the application to special relativity, another important source of motivation for the study of K-loops is the problem of existence of a proper neardomain. This question is closely related with the structure of sharply 2-transitive groups and is still open. Frobenius groups with many involutions seem to be a reasonable generalization of sharply 2-transitive groups. Examples of these are constructed and studied.

Let's now go into more detail: A set L with a binary operation $\cdot : L \times L \to L$ is called a *quasigroup* if the equations $ax = b$ and $ya = b$ have unique solutions x, y in L for all $a, b \in L$. Note that we use the convention $ab := a \cdot b$. If the quasigroup L contains an identity element, usually denoted by 1, then L is called a *loop*. Loops which satisfy the Bol condition,

$$a(b \cdot ac) = (a \cdot ba)c \quad \text{for all } a, b, c \in L,$$

are called *Bol loops*. Here and later we use the *dot-convention*[1] introduced in section 2. A *K-loop* is a Bol loop L with the *automorphic inverse property*

$$(ab)^{-1} = a^{-1}b^{-1} \quad \text{for all } a, b \in L.$$

We'll show later that it makes sense to use inverses in Bol loops.

The principal aim of this book is the development of theory and applications of K-loops from scratch, and the presentation of examples.

---

[1] Undefined expressions used in the introduction can be looked up in the text.

If "·" is a binary operation on L, then so is the *dual operation*

$$* : L \times L \to L; \; (a, b) \mapsto a * b := ba.$$

We shall speak of the dual quasigroup, loop, etc. The *dual of an identity*[2] is this same identity in $(L, *)$, i.e., "$*$" replaces "·", rewritten into an identity in $(L, \cdot)$. For example, the dual of the *Bol identity*, the *right Bol identity*, reads

$$(ca \cdot b)a = c(ab \cdot a).$$

A binary operation is called *commutative* if $\cdot = *$. This implies of course that every identity is equivalent to its dual.

The class of Bol loops is not self-dual, i.e., the Bol identity does not imply the right Bol identity. This can be shown by example easily. Therefore we have to make a decision between two options. We choose the one which is most easily expressed with the left multiplication maps. E.g., we use the (left) Bol identity, the left inverse property, the left alternative law, etc. This fits nicely with mappings applied on the left, such as $f(x)$. The alternative would be to dualize everything. Of course, there is no essential difference between the two possibilities.

This work is subdivided roughly into two parts. The first part tries to give a coherent exposition of the presently known theory. Much of the material has been known one way or another, but it is scattered rather widely through the literature.

In §1 we collect a few results used later in the text. The main subsections E, F, G, are a preparation for the examples constructed from the classical groups in §9. Some of these results seem to be

---

[2] The word "identity" here refers to an expression which is to be true for all choices of the variables. Admittedly, this is a bit fuzzy, but it is made precise in any textbook on universal algebra, e.g., Ihringer: Allgemeine Algebra, Teubner, Stuttgart, 1988. Such identities are not to be confused with the identity element (of a loop for instance), to which we frequently refer simple as an identity. The context will always make the meaning clear.

new. Therefore they are proved in the text. Subsections D and H give results used in §11. Subsections A, B, C serve merely to fix notation, and to provide some statements for convenient reference.

Transversals pop up throughout the development. Every time we discuss some new axioms, we check what this means in terms of transversals. §2 sets the stage by showing how a transversal always has the structure of a left loop. And conversely, every left loop can be seen as a transversal inside its left multiplication group with respect to the left inner mapping group. Transassociants and the quasidirect product give a more flexible tool to realize left loops as transversals. In particular, the left multiplication group can be identified as a quasidirect product. Another important application is presented in §7, where we describe certain infinite Frobenius groups using the quasidirect product.

§3 contains basic facts about left loops with the left inverse property. We discuss the impact of the automorphic inverse property, and of the A_ℓ-property, i.e., when all precession maps are automorphisms. In particular, without using Bol, we derive various identities for the precession maps, which were known to hold for K-loops. These considerations lead to the notion of a Kikkawa loop, which is also treated in this section. Finally, we show that a left loop with the Bol condition is in fact a loop.

In the following §4 we introduce some isotopy theory, just as much as we need, plus BELOUSOV's theorem characterizing G-loops, i.e., loops such that all loop isotopes are isomorphic.

In §5 we discuss the nuclei, and we prove $\mathcal{N}_\ell = \mathcal{N}_m$ for left loops with the left inverse property. As a consequence we obtain that the center, left and middle nucleus coincide for Kikkawa loops, and in particular for K-loops.

After a short introduction of the left power alternative property, in §6 we characterize Bol loops, among others, as loops with all isotopes satisfying the left inverse property, and as loops such that all isotopes are left alternative. In particular, every loop isotope of a Bol loop is a Bol loop.

We are then ready to prove KREUZER's theorem, stating that K-loops are A_ℓ-loops. As a consequence various characterizations of K-loops are given, the most striking being (6.9) and (6.11). They imply that a left loop is already a K-loop if it is a 2-divisible Kikkawa loop, or if it is a uniquely 2-divisible, left power alternative A_ℓ-left-loop with a fixed point free, involutory automorphism. We also present (without proof) ROBINSON's results on the isotopes of K-loops.

We conclude this section by describing BRUCK's construction of K-loops from uniquely 2-divisible groups (and even Moufang loops), and GLAUBERMAN's half embedding.

"Frobenius groups with many involutions", introduced in §7, generalize the notion of a sharply 2-transitive group, and so does the more restrictive notion of a "specific group". A Frobenius group with many involutions can be realized as a quasidirect product of a K-loop L and an appropriate transassociant T. In fact, for this construction to work, it is necessary and sufficient that the left inner mapping group $\mathcal{D}(L)$ acts fixed point free on $L^\#$. The point is: There do exist examples which are not semidirect products. §9 and §11 provide such examples, i.e., non-associative K-loops with fixed point free $\mathcal{D}(L)$. These would yield a proper neardomain if the inner mapping group could be enlarged to a fixed point free transassociant, which acts transitively, see §7.D, in particular (7.9). Such an example doesn't seem to be known at present. We hope that the study of examples of K-loops with fixed point free left inner mappings eventually can pave the way to deciding the question whether proper neardomains exist. To make the context clear, neardomains and their connection with sharply 2-transitive groups are introduced in subsection D of §7.

We'll also construct some specific groups of characteristic 0. For (odd) prime characteristic, we can only refer to GABRIEL's examples from the Burnside groups.

In §8 we give a short introduction into ZIZIOLI's theory of fibered loops and their relation with geometry.

The second part is devoted to examples, the heart of every theory. In §9 we construct transversals in the classical groups over appropriate ordered fields. They are all K-loops. We can compute the left inner mapping groups and the centers in case of the special and general linear groups. We can extract the examples with fixed point free left inner mapping groups in these cases, and for the pseudo-orthogonal and pseudo-unitary groups. Because the Lorentz group of special relativity is (a subgroup of) the group $O(3,1)$, we study the pseudo-orthogonal and pseudo-unitary groups in some detail. Finally, we give a purely algebraic construction of an example of GABRIELI and KARZEL for a fibered K-loop with non-abelian fibers.

In §10 we apply the results from the previous section to give a rather smooth derivation of the formula for the relativistic velocity addition. We do this by using a natural one-to-one correspondence between velocities of frames and Lorentz boosts, and the fact that the Lorentz boosts form a transversal inside the Lorentz group, which is a K-loop.

An interesting construction of KOLB and KREUZER, presented in §11, also yields K-loops with fixed point free transassociant. Indeed, we get plenty of examples of such K-loops of exponent 2, and also K-loops that lead to specific groups of characteristic 0. Moreover, these K-loops admit an Aut-invariant fibration.

Derivations, as introduced in §12, are a method which modifies the multiplication in groups. It turned out useful to constructing various counterexamples. This method is well-known to constructing quasifields from fields, see [62] and [63]. In particular, Bol-quasifields have been found this way [21, 23]. It has also been used, probably unconsciously, i.e., without axiomatizing the method, for the construction of loops.

The appendix highlights some lines of development in the history of the subject.

1. Preliminaries

In this section we collect some definitions, notations, and results for the convenience of the reader. We'll assume that the reader is familiar with standard knowledge in algebra. All that we use without reference can be found in [84]. The major part of this section is devoted to some preparation for §9. The subjects are ordered fields, hermitian matrices, and general matrices. Lack of reference to a published source is not a claim for originality. However, some of the results seem to be new.

The set of integers will be denoted by $\mathbf{Z}$, the set of positive integers (natural numbers) by $\mathbf{N}$, and the set of rationals by $\mathbf{Q}$. The cardinality of a set S will be denoted by $|S|$.

A. GROUPS

Despite the fact that loops generalize groups, we will not build up group theory for obvious reasons. Instead, we'll assume basic notions from group theory, as they can be found for instance in [5, 42, 47].

Let G be a group. The identity will mostly be denoted by 1. We'll write $|g|$ for the *order* of an element $g \in G$. Whenever it makes sense this notion will also be used for loops. An element of order 2 is called an *involution*. For every subset S of G we put $S^{\#} := S \setminus \{1\}$. The subgroup generated by S will be denoted as $\langle S \rangle$. The *centralizer* of S in U, where U is a subgroup of G,

$$C_U(S) := \{g \in U; \forall x \in S : gxg^{-1} = x\}$$

is a subgroup of U. The *kernel* of a homomorphism η will be denoted by $\ker \eta$. The inner automorphism given by $g \in G$ will be written as

$$\hat{g} : \begin{cases} G \to G \\ x \mapsto gxg^{-1} \end{cases}.$$

The map $g \mapsto \hat{g}$ is an epimorphism $G \to \operatorname{Inn} G$, the *inner auto-morphism group* of G. Its kernel is the *center* of G, denoted by $\mathcal{Z}(G)$.

For $a \in G$ put $[a, b] := a^{-1}b^{-1}ab$, the *commutator* of a and b. Let $S, T \subseteq G$, then[1]

$$[S, T] := \langle [s, t]; \ s \in S, \ t \in T \rangle .$$

Note that $[S, T]$ is a group by definition. From [61; 2.1.5, p. 33] one derives

(1.1) *Let G be a group and $S \subseteq G$ with $S^{-1} \subseteq S$. If $\big[[a, b], c\big] = 1$ for all $a, b, c \in S$, then $\big[[S, S], S\big] = \{1\}$.* ∎

A natural number n is called an *exponent* of the group G if $g^n = 1$ for all $g \in G$. This notion will later be used for left power alternative loops, as well.

(1.2) *A group of exponent 2 is abelian.* ∎

If G and U are abelian groups, then the set of all homomorphisms $G \to U$ is an abelian group, denoted by $\operatorname{Hom}(G, U)$. Its zero element is the zero map denoted by $\mathbf{0} : G \to U; \ a \mapsto 0$. Notice that we do not distinguish the zero maps with respect to domain and range.

B. PERMUTATION GROUPS

Let P, S be sets. P^S will be the set of all maps $S \to P$. The *symmetric group* on P will be denoted by $\mathcal{S}_P$. It is the set of all bijective elements of P^P (also called *permutations* of P) with the usual composition of functions. Throughout this work $\mathbf{1}_P$ denotes the identity map of P, which is abbreviated to $\mathbf{1}$ if the set is clear from the context. $\mathbf{1}$ is the identity element of $\mathcal{S}_P$.

A homomorphism π of a group G into $\mathcal{S}_P$ will be called a *permutation representation*. This is the same to say that G acts on P.

---

[1] Here and later we'll drop the extra braces, writing briefly $\langle \ldots \rangle$ rather than $\langle \{ \ldots \} \rangle$.

The representation will be called

faithful if $\ker \pi$ is trivial;

(sharply) k-transitive ($k \in \mathbf{N}$) if for every k distinct elements $a_1, \ldots, a_k$ and $b_1, \ldots, b_k$ from P there exists (exactly one) $g \in G$ such that $\pi(g)(a_i) = b_i, i \in \{1, \ldots, k\}$;

transitive if it is 1-transitive;

regular if it is sharply 1-transitive.

The notion of (sharply) transitive action makes sense for subsets L of G, too. So we can speak of L acting regularly, or transitively on P if the defining condition from above holds with L replacing G.

We'll frequently omit the π in $\pi(g)(a)$, writing briefly $g(a)$. In most cases we need not even specify π. We then simply write: (G, P) is a *permutation group*. Note that this is stretching common usage quite a bit, because we do not assume G to be a subgroup of $\mathcal{S}_P$, nor do we require the representation to be faithful.

The elements of P will often be referred to as *points*. Let S be a subset of G. The set

$$\mathrm{Fix}(S) := \{p \in P; \ \forall g \in S : g(p) = p\}$$

will be called the set of fixed points of S. We'll say S acts *fixed point free* if $\mathrm{Fix}(\{g\}) = \varnothing$ for all $g \in S^\#$, i.e., no non-identity element of S has a fixed point. If P is a group (or a groupoid) with identity 1 and S consists of elements which fix 1 by definition, e.g., automorphisms, then S is called *fixed point free* if it acts fixed point free (in the old sense) on $P^\#$. We'd like to emphasize that $S := \{\mathbf{1}\}$ is considered fixed point free in both cases.

Let G, G' be groups acting on sets P, P', respectively. The representations will be called *equivalent* if there exists a bijection $\mu : P \to P'$ and an isomorphism $\alpha : G \to G'$ such that $\alpha(g)(x) = \mu g \mu^{-1}(x)$ for all $g \in G, x \in P'$. The pair of mappings (α, μ) will be called an *equivalence*. The situation is illustrated by the diagram.

$$
\begin{array}{ccc}
P & \xrightarrow{\ g\ } & P \\
\downarrow \mu & & \downarrow \mu \\
P' & \xrightarrow{\alpha(g)} & P'
\end{array}
$$

If the representation is faithful, then α is uniquely determined by μ. This will be expressed sometimes by saying "μ induces an equivalence".

Transitive actions can be characterized intrinsically up to equivalence. This is taken from [5; (5.7), (5.8), (5.9), p. 15f].

(1.3) *Let (G, P) be a transitive permutation group, and let Ω be the stabilizer of a point $e \in P$.*

(1) *The kernel N of the representation is $N = \bigcap_{g \in G} g\Omega g^{-1}$. This is the largest normal subgroup of G contained in Ω.*

(2) *G acts transitively on the set of left cosets G/Ω by left multiplication. This action is equivalent to the action of G on P.*

(3) *The stabilizer of $a \in P$ is $g\Omega g^{-1}$ for every $g \in G$ with $g(e) = a$. Therefore all stabilizers of elements in P are conjugate.*   ∎

The representation of G on G/Ω by left multiplication is called the *natural permutation representation* of G on G/Ω. The largest normal subgroup N of G inside Ω is called the *core* of Ω in G. It is the kernel of the natural permutation representation of G on G/Ω as in (2). A subgroup Ω of G will be called *corefree* if $N = \{1\}$, i.e., if the core is trivial.

C. GEOMETRY

An *incidence structure* is a pair of sets $(P, \mathcal{L})$ such that $\mathcal{L}$ consists of subsets of P. Moreover, we require that $|A| \geq 2$ for all $A \in \mathcal{L}$. The elements of P will be called *points*.

An incidence structure is called an *incidence space* (or *linear space*) if for every $a, b \in P$, $a \neq b$, there exists exactly one $A \in \mathcal{L}$ with $a, b \in A$. Then A is called the *line* passing through a and b. Similar geometric language will be used freely.

Let $(P, \mathcal{L})$, $(P', \mathcal{L}')$ be incidence spaces. A bijective map $\alpha : P \to P'$ will be called an *isomorphism* if $\alpha(A) \in \mathcal{L}'$ for all $A \in \mathcal{L}$, i.e., lines are mapped to lines. As usual α is called an *automorphism* if $P = P'$. The set of all automorphisms of P is denoted by $\mathrm{Aut}(P, \mathcal{L})$, and is clearly a group.

A binary relation $\|$ on $\mathcal{L}$ is called a *parallelism* if $\|$ is an equivalence relation, and for all $a \in P$, $A \in \mathcal{L}$ there exists a unique

$B \in \mathcal{L}$ with $A \parallel B$ and $a \in B$. The triple $(P, \mathcal{L}, \parallel)$ will then be referred to as an *incidence space with parallelism*.

Automorphisms of an incidence space with parallelism are automorphisms α of the incidence space $(P, \mathcal{L})$ that preserve $\parallel$, more precisely: if $A \parallel B$, then $\alpha(A) \parallel \alpha(B)$, for $A, B \in \mathcal{L}$. The set of such automorphisms is a subgroup of $\mathrm{Aut}(P, \mathcal{L})$, denoted by $\mathrm{Aut}(P, \mathcal{L}, \parallel)$. An automorphism α of $(P, \mathcal{L})$ is called a *dilatation* if $\alpha(A) \parallel A$ for all $A \in \mathcal{L}$. Clearly, the set of all dilatations is a subgroup of $\mathrm{Aut}(P, \mathcal{L}, \parallel)$.

The following is folklore. For completeness we recall the proof.

(1.4) Let $(P, \mathcal{L}, \parallel)$ be an incidence space with parallelism. Assume that $P \notin \mathcal{L}$, i.e., P is not a line. If a dilatation α has two fixed points, then $\alpha = 1$.

Proof. Let a, b be fixed points of the dilatation α. Denote the line joining a, b by A. For $c \in P \setminus A$ we have that the lines joining a with c, and joining b with c are both fixed. Then so is their intersection c. Hence every point in $P \setminus A$ is fixed. Applying the same line of reasoning to a and c completes the proof. ∎

D. BINOMIAL COEFFICIENTS

In §11 we'll need a corollary to a theorem of LUCAS on binomial coefficients. Recall that for all $n, k \in \mathbf{N} \cup \{0\}$

$$\binom{n}{k} = 0 \iff k > n.$$

(1.5) Theorem. Let p be a prime, $n, n', k, k' \in \mathbf{N} \cup \{0\}$, and $n', k' < p$ then

$$\binom{np + n'}{kp + k'} \equiv \binom{n}{k}\binom{n'}{k'} \mod p.$$

Proof. See [32] for a short proof. ∎

Apply this repeatedly to obtain

(1.6) Let p be a prime, $n \in \mathbf{N}$, and choose $r \in \mathbf{N} \cup \{0\}$ maximal such that p^r divides n, then $\binom{n}{p^r} \not\equiv 0 \mod p$. ∎

E. Ordered Fields

Let R be a field. As usual, put $R^* := R \setminus \{0\}$, the multiplicative group of R. A subset $P \subseteq R$ is called an *ordering* if for all $x \in R$ we have $x \in P$, or $-x \in P$, or $x = 0$, and these conditions are mutually exclusive. Moreover, we require that $P + P \subseteq P$, and $PP \subseteq P$. Therefore P is a additively and multiplicatively closed subset of R^* such that R^* is the disjoint union of P and $-P$. These conditions allow us to define a relation $<$ on R by

$$x < y : \Longleftrightarrow y - x \in P \quad \text{for all } x, y \in R.$$

This relation has the usual properties for inequalities, see [84; Ch. XI, §1] for more. A field is called *ordered* if it possesses an ordering. In this case we will tacitly use the relation $<$ and also $\leq$, which is defined in the usual way, and which is an order relation. Notice that squares and also sums of squares are always ≥ 0.

In an ordered field R, -1 is not a square, thus there exists a quadratic field extension $K := R(i)$, where $i^2 = -1$. K has a unique non-identity automorphism $z \mapsto \bar{z}$, which fixes R element-wise and maps i to $-i$. We'll use this notation whenever we deal with ordered fields.

An ordered field R is called *pythagorean* if for every $a \in R$ the element $1 + a^2$ is a square in R. Hence every sum of squares is a square. Note that we deviate from standard usage, as we do require a pythagorean field to be ordered. This is more convenient, since we do not deal with non-ordered "pythagorean" fields.

If a is a square in R we write $\sqrt{a}$ for the unique positive square root of a in R.

In an ordered field, we can define $|a|$ for $a \in R$ as usual, and we obtain $|a| = \sqrt{a^2}$. When R is pythagorean, we can and will also write $|a| := \sqrt{a\bar{a}} \in R$ (!) for $a \in K$. This should not be confused with the order of an element in a group, or the cardinality of a set. The context will always make the meaning clear.

A field is called *euclidean* if the set of squares is an ordering, i.e., the field is pythagorean and each non-zero element is either a square or the negative of a square. Such fields have a unique ordering.

An ordered field R is called *real closed* if $R(i)$ is algebraically closed. These are exactly those ordered fields which do not have an ordered algebraic extension, see [84; Ch. XI, §2]. Note that real closed fields are euclidean, hence have a unique ordering.

An ordered extension field R' of R is called a *real closure* of R, if the extension is algebraic, and if R' is real closed. The intersection of all real closures of R in a fixed algebraic closure $\bar{R}$ will be denoted by R^{tot}. The ordered field R is called *totally real* if $R = R^{\text{tot}}$.

For later use we'll show

(1.7) $\mathbf{Q}^{\text{tot}}$ is totally real, but not euclidean. Every euclidean closure of $\mathbf{Q}$ is a euclidean field, which is not totally real.

Proof. Since $\mathbf{Q}^{\text{tot}}$ is pythagorean, $\sqrt{2}$ is an element. Now $\mathbf{Q}(\sqrt{2})$ has two different orderings, one with $\sqrt{2}$ positive, one with $\sqrt{2}$ negative. Let R_1 and R_2 be real closures of $\mathbf{Q}(\sqrt{2})$ with respect to these orderings.[2] R_1 and R_2 induce two different orderings on $\mathbf{Q}^{\text{tot}}$, hence $\mathbf{Q}^{\text{tot}}$ cannot be euclidean.

For the second assertion, let E be a euclidean closure of $\mathbf{Q}$. It follows from the construction in the proof of [9; Satz 5] that every element in E has degree a power of 2 over $\mathbf{Q}$. Let a be a root of an irreducible polynomial f of odd degree over $\mathbf{Q}$ such that all roots of f are in $\mathbf{Q}(a)$, e.g., $f(x) = x^3 - 3x + 1$.[3] Let R be a real closure of $\mathbf{Q}$ in a fixed algebraic closure $\bar{\mathbf{Q}}$ of $\mathbf{Q}$. Every odd degree polynomial such as f has a root in R [120; §71, Satz 2, p. 239], hence f splits over R. Therefore $a \in \mathbf{Q}^{\text{tot}}$. Now, if E were totally real it would contain $\mathbf{Q}^{\text{tot}}$, which is impossible since $a \notin E$. ∎

F. HERMITIAN MATRICES AND THE POLAR DECOMPOSITION

We continue to use the symbols R and $K = R(i)$ as in the preceding subsection. Let $n \geq 2$ be a fixed integer. The set of $m \times n$-matrices over K (or any other ring) will be denoted by

---

2 For the existence see [84; Ch. XI.2.11, p. 397].

3 If a is a root, then $a^2 - 2$ and $-a^2 - a + 2$ are the other roots.

$K^{m \times n}$, $m \in \mathbf{N}$. For $X \in K^{m \times n}$ let X^{T} denote the transpose of X. Then $X^{\mathrm{T}} \in K^{n \times m}$. The automorphisms of K extend to $K^{m \times n}$ by componentwise application. Let $X^* := \overline{X}^{\mathrm{T}}$ for every $X \in K^{m \times n}$. We'll use the common shorthand $K^n := K^{n \times 1}$, so the elements of K^n are column vectors.

Remark. The symbol * which has been just introduced to denote the hermitian conjugate, also denotes "field without 0". This should not cause any trouble. Care is needed only here and in §9, when we use hermitian matrices to construct K-loops.

(1.8) *Let R be a pythagorean field, and $n \in \mathbf{N}$. Then the map*

$$\|\cdot\| : \begin{cases} R^n \to R \\ \mathbf{v} \mapsto \|\mathbf{v}\| := \sqrt{\mathbf{v}^{\mathrm{T}}\mathbf{v}} \end{cases}$$

is a norm, and we have $|\mathbf{v}^{\mathrm{T}}\mathbf{w}| \leq \|\mathbf{v}\|\|\mathbf{w}\|$.

Proof. $\mathbf{v}^{\mathrm{T}}\mathbf{v}$ is a sum of squares, hence a square in R. Therefore $\|\cdot\|$ is well-defined. Clearly, $\|\cdot\|$ is induced by a positive definite bilinear form, therefore it is a norm. The inequality is the well-known "Schwarz inequality". It is proved in [84; XIV.7, p. 510].[4] ∎

Consider the set $\mathcal{H}$ of positive definite hermitian $n \times n$-matrices over K

$$\begin{aligned} \mathcal{H} &= \mathcal{H}(n, K) \\ &= \{A \in K^{n \times n}; \ A = A^*, \forall \mathbf{v} \in K^n \setminus \{0\} : \mathbf{v}^*A\mathbf{v} > 0\}, \end{aligned}$$

and write

$$\mathrm{U}(n, K) = \{U \in K^{n \times n}; \ UU^* = I_n\}$$

for the corresponding unitary group. Here I_n denotes the $n \times n$ identity matrix. R will be called n-*real* if the characteristic polynomial of every matrix in $\mathcal{H}$ splits over K into linear factors. In fact, it is well-known and easy to see that in this case the characteristic polynomial already splits over R. Therefore, we can always

---

[4] The given reference does not state the Schwarz inequality in the generality needed, but the proof carries over verbatim. It is, by the way, the well-known standard proof from calculus.

assume that all eigenvalues of elements of $\mathcal{H}$ are "real"(i.e., are elements of R). Clearly, n-real implies $(n-1)$-real.

By [122; Prop. 7] (see also [89]), we have

(1.9) R is n-real for all $n \in \mathbf{N}$ if and only if R is totally real. ∎

Remarks 1. In fact, both [122, 89] prove this theorem only for symmetric matrices over R. However, the proofs given in [89] apply after minor modification to the hermitian case as they stand. This is true in particular for [89; Theorem 2] and the Corollary and Remarks following it.

2. The class of totally real fields strictly contains the class of real closed fields and the class of hereditary euclidean fields (cf. [122]). Moreover, the intersection of totally real fields (all contained in a common extension field) is totally real.

(1.10) Theorem. R is 2-real if and only if R is pythagorean. For $n \geq 2$ every n-real field is pythagorean.

The *Proof* is an adaption of the proof of [89; Lemma 1] to our situation.

If R is 2-real, then let $a \in R$ be positive. The matrix

$$A := \begin{pmatrix} a+2 & a \\ a & a \end{pmatrix}$$

is clearly hermitian over R (it is in fact symmetric). The eigenvalues are

$$a + 1 \pm \sqrt{1 + a^2},$$

which are to be in R, since A is clearly positive definite. Hence $1 + a^2$ is a square in R, and because a was arbitrary, we conclude that R is pythagorean.

Conversely, assume that R is a pythagorean field. A general hermitian 2×2-matrix A has the form

$$A = \begin{pmatrix} \alpha & a \\ \bar{a} & \beta \end{pmatrix},$$

with $\alpha, \beta \in R$ and $a \in R(i)$. The characteristic polynomial of A is

$$f(x) := x^2 - (\alpha + \beta)x + \alpha\beta - a\bar{a}$$

and has discriminant $(\alpha + \beta)^2 - 4(\alpha\beta - a\bar{a}) = (\alpha - \beta)^2 + 4a\bar{a}$. This is a sum of squares in R, hence a square. Thus f splits over R, and R is 2-real.

The last assertion follows from the fact that n-real fields are 2-real.   ∎

(1.11) $I_n \in \mathcal{H}$, $\mathcal{H}^{-1} = \mathcal{H}$ and $A^*\mathcal{H}A = \mathcal{H}$ for all $A \in \mathrm{GL}(n, K)$.

Proof. The first statement is clear. Take $B \in \mathcal{H}$. We have $(B^{-1})^* = (B^*)^{-1} = B^{-1}$. Furthermore, for every non-zero $\mathbf{v} \in K^n$ holds

$$0 < (B^{-1}\mathbf{v})^* B(B^{-1}\mathbf{v}) = \mathbf{v}^* B^{-1} B B^{-1} \mathbf{v} = \mathbf{v}^* B^{-1} \mathbf{v},$$

hence B^{-1} is positive definite.[5]

Moreover,

$$(A^* B A)^* = A^* B A, \quad \text{and} \quad \mathbf{v}^* A^* B A \mathbf{v} = (A\mathbf{v})^* B(A\mathbf{v}) > 0,$$

for all $\mathbf{v} \in K^n \setminus \{0\}$. Hence $A^* B A$ is positive definite.   ∎

We will have use for the spectral theorem [84; XIV.12.4]. Note that our assumption for K makes sure that this theorem is true.

(1.12) Theorem. *Let R be n-real and let $A \in \mathcal{H}$. The eigenvalues of A are in R, and there exists a $U \in \mathrm{U}(n, K)$ such that UAU^{-1} is diagonal. This means, there exists an orthonormal basis of K^n, consisting of eigenvectors of A. If $A \in \mathrm{GL}(n, R)$, then U can be taken from $\mathrm{GL}(n, R)$. Furthermore, it is possible to chose U such that $\det U = 1$.*

Proof of the last statement. If $\det U = \alpha$, then $\alpha\bar{\alpha} = 1$ (since $\det U^* = \overline{\det U}$). Hence we can replace the first column $\mathbf{c}$, say, of U by $\alpha^{-1}\mathbf{c}$ to obtain a matrix U'. It is easy to see that U' is unitary. Clearly, $\det U' = 1$, hence the result.   ∎

(1.13) *Let R be n-real. For any $A \in K^{n \times m}, m \in \mathbf{N}$, the eigenvalues of AA^* are squares in R.*

Proof. Let λ be an eigenvalue of AA^* with eigenvector $\mathbf{v} \in K^n$, i.e., $AA^*\mathbf{v} = \lambda\mathbf{v}$. It follows that

$$\mathbf{v}^* AA^* \mathbf{v} = \lambda \mathbf{v}^* \mathbf{v} \quad \text{and} \quad \lambda = \frac{1}{(\mathbf{v}^*\mathbf{v})^2}(\mathbf{v}^*\mathbf{v})\big((A^*\mathbf{v})^*(A^*\mathbf{v})\big)$$

---

[5] An alternative argument uses eigenvalues.

which is a sum of squares in R. Since R is pythagorean by (1.10), we conclude that λ is a square in R.   ∎

(1.14) *The map* $\kappa : \mathcal{H} \to \mathcal{H}; X \mapsto X^2$ *is injective.*

Proof. $X^2 = X^* I_n X \in \mathcal{H}$, by (1.11). Now assume that $A^2 = B^2$ for $A, B \in \mathcal{H}$. There is no loss in generality to assume that R is n-real (or even real closed), since otherwise we pass to an extension field of R.

By the spectral theorem (1.12) we can decompose K^n into an orthogonal sum of eigenspaces $V_1, \ldots, V_r$ of A, say. Denote the corresponding (distinct!) eigenvalues by $\lambda_1, \ldots, \lambda_r$. Thus $A\mathbf{v} = \lambda_k \mathbf{v}$ and $A^2 \mathbf{v} = \lambda_k^2 \mathbf{v} = B^2 \mathbf{v}$ for all $\mathbf{v} \in V_k, k \in \{1, \ldots, r\}$. Note that $\lambda_k > 0$ for all $k \in \{1, \ldots, r\}$. Applying the same argument to B, we conclude that the V_k are the eigenspaces of B, as well. Since B is positive definite, the λ_k are the eigenvalues of B. This entails $A = B$, and the injectivity of κ.   ∎

(1.15) *Let R be n-real and take $A, B \in \mathcal{H}$, then $A \in B\,\mathrm{U}(n, K)$ if and only if $A = B$, i.e., every coset of $\mathrm{U}(n, K)$ in $\mathrm{GL}(n, K)$ contains at most one element of $\mathcal{H}$.*

Proof. Let $A = \mathrm{diag}(\lambda_1, \ldots, \lambda_n)$ and $C := B^{-1} = (\, \mathbf{c}_1 \quad \cdots \quad \mathbf{c}_n \,)$, where $\mathbf{c}_k \in K^n$ are the columns of C. Note that $C \in \mathcal{H}$ by (1.11). We compute $CA = (\, \lambda_1 \mathbf{c}_1 \quad \cdots \quad \lambda_n \mathbf{c}_n \,) \in \mathrm{U}(n, K)$. Using $\lambda_k \in R$, for $k \neq j$ we have

$$0 = (\lambda_k \mathbf{c}_k)^* (\lambda_j \mathbf{c}_j) = (\lambda_k \lambda_j) \mathbf{c}_k^* \mathbf{c}_j.$$

Since $\lambda_k \lambda_j \neq 0$ we infer that C has orthogonal columns, i.e.,

$$C^2 = C^* C = \mathrm{diag}(\mathbf{c}_1^* \mathbf{c}_1, \ldots, \mathbf{c}_n^* \mathbf{c}_n).$$

From (1.14) and (1.10) we conclude $C = \mathrm{diag}(\mu_1, \ldots, \mu_n)$ for appropriate $\mu_k \in R$. Thus $CA = \mathrm{diag}(\lambda_1 \mu_1, \ldots, \lambda_n \mu_n) \in \mathrm{U}(n, K)$. Since $\lambda_k \mu_k > 0$ we must have $B^{-1} A = CA = I_n$, as claimed.

The general case where A is not necessarily diagonal can be reduced to the case discussed above by (1.12).   ∎

Let $A \in \mathrm{GL}(n, K)$. A factorization $A = BU$, with $B \in \mathcal{H}, U \in \mathrm{U}(n, K)$ is called *polar decomposition*. We have just seen that the polar decomposition is unique if it exists. The existence of the polar decomposition is the subject of the next lemma. It is the key to the construction of a big class of K-loops in §9.

(1.16) *Let G be a subgroup of $\mathrm{GL}(n,K)$ and put*

$$L_G := G \cap \mathcal{H}(n,K), \quad \Omega_G := G \cap \mathrm{U}(n,K).$$

The following are equivalent.

   (I) $G = L_G \Omega_G$;

   (II) $AA^* \in \kappa(L_G)$, *for all* $A \in G$.

Proof. (I) $\Longrightarrow$ (II): Let $A = BU \in G, B \in L_G, U \in \Omega_G$. We have $AA^* = BUU^*B^* = BB^* = B^2 \in \kappa(L_G)$.

(II) $\Longrightarrow$ (I): For $A \in G$ let $B \in L_G$ be subject to $B^2 = AA^*$. Now $U := B^{-1}A$ is in Ω, because $UU^* = B^{-1}AA^*(B^{-1})^* = B^{-1}B^2B^{-1} = I_n$. Hence $G \subseteq L_G\Omega_G$. ∎

Remark. A very special case of this is contained in [49; (1.1.e)].

G. Miscellaneous Results for Matrices

We'll have to extract square roots from certain hermitian matrices several times.

(1.17) *Let R be n-real, and let $A_1,\dots,A_k \in \mathcal{H}(n,K)$ such that every eigenvalue of every A_j is a square in R. Then there exists a polynomial $f \in R[x]$ with $f(A_j) \in \mathcal{H}(n,K)$ and $\big(f(A_j)\big)^2 = A_j$ for all $j \in \{1,\dots,k\}$.*

Proof. Let Λ be the set of all the eigenvalues of all the A_j. Clearly, Λ is a finite set, and by assumption every element $\lambda \in \Lambda$ has a unique positive square root $\sqrt{\lambda} \in R$. By LAGRANGE's interpolation formula [120; §25] there exists a polynomial $f \in R[x]$ with $f(\lambda) = \sqrt{\lambda}$ for all $\lambda \in \Lambda$. We'll show that f does the job.

Let $j \in \{1,\dots,k\}$. By (1.12) there exists $U \in \mathrm{U}(n,K)$ such that

$$A_j = U \operatorname{diag}(\lambda_1,\dots,\lambda_n)U^{-1}, \text{ with } \lambda_\ell \in \Lambda \text{ for all } \ell \in \{1,\dots,n\},$$

because R is n-real. Thus we can compute

$$f(A_j) = f\big(U \operatorname{diag}(\lambda_1, \ldots, \lambda_n)U^{-1}\big) = Uf\big(\operatorname{diag}(\lambda_1, \ldots, \lambda_n)\big)U^{-1}$$
$$= U \operatorname{diag}\big(f(\lambda_1), \ldots, f(\lambda_n)\big)U^{-1} = U \operatorname{diag}(\sqrt{\lambda_1}, \ldots, \sqrt{\lambda_n})U^{-1}.$$

This shows the result.   ∎

The following simple lemma is taken from [96; §39, p. 169].

(1.18) *Let* $A = \operatorname{diag}(\lambda_1, \ldots, \lambda_n)$, $B \in K^{n \times n}$ *with* $|\{\lambda_1, \ldots, \lambda_n\}| = n$ *(i.e., all* λ_k *distinct), then we have* $AB = BA \iff B = \operatorname{diag}(\alpha_1, \ldots, \alpha_n)$ *for some* $\alpha_k \in K$.   ∎

For $1 \le k \le n - 1$, we consider the sequence of embeddings

$$\Delta_k : \operatorname{GL}(2, K) \to \operatorname{GL}(n, K); \ A \mapsto \begin{pmatrix} I_{k-1} & & \\ & A & \\ & & I_{n-1-k} \end{pmatrix},$$

where the images are the block diagonal matrices with A placed in the (k, k)-position. Recall that $\operatorname{O}(n, R) = \{U \in R^{n \times n}; \ UU^{\mathrm{T}} = I_n\} = \operatorname{U}(n, K) \cap \operatorname{GL}(n, R)$ is the *orthogonal group*. $\operatorname{SU}(n, K)$ and $\operatorname{SO}(n, R)$ refer to corresponding "special" groups, consisting of the elements of $\operatorname{U}(n, K)$, $\operatorname{O}(n, R)$, respectively, of determinant 1.

First a few trivial observations.

(1.19) *For all* $k \in \{1, \ldots, n - 1\}$ *the map* Δ_k *is a monomorphism, and we have*

(1) $\Delta_k\big(\operatorname{SU}(2, K)\big) \subseteq \operatorname{SU}(n, K)$;

(2) $\Delta_k\big(\operatorname{SL}(2, R)\big) \subseteq \operatorname{SL}(n, R)$;

(3) $\Delta_k\big(\operatorname{SO}(2, R)\big) \subseteq \operatorname{SO}(n, R)$;

(4) $\Delta_k\big(\mathcal{H}(2, K)\big) \subseteq \mathcal{H}(n, K)$.   ∎

We now look at a set of generators for $\operatorname{SU}(n, K)$ and $\operatorname{SO}(n, R)$ which might be of independent interest. We shall prove

(1.20) Theorem. $\operatorname{SU}(n, K) = \big\langle \bigcup_{k=1}^{n-1} \Delta_k\big(\operatorname{SU}(2, K)\big) \big\rangle$ *and*

$\operatorname{SO}(n, R) = \big\langle \bigcup_{k=1}^{n-1} \Delta_k\big(\operatorname{SO}(2, R)\big) \big\rangle$.

The proof must be postponed after the following lemma. It will be carried out only for $\operatorname{SU}(n, K)$. In particular, an analogue of

the following lemma for $SO(n, R)$ holds. We remark in advance
that the proofs for $SO(n, R)$ are virtually identical. We denote the
standard basis of K^n by $e_1, \ldots, e_n$, e.g., $e_1 := (1, 0, \ldots, 0)^T$.

(1.21) $S := \left\langle \bigcup_{k=1}^{n-1} \Delta_k(SU(2, K)) \right\rangle$ *acts transitively on the set of*
normalized vectors $v \in K^n$ *(i.e.,* v *such that* $\|v\|^2 = v^* v = 1$ *).*

Proof. Let $v := (v_1, \ldots, v_n)^T \in K^n$ with $v^* v = |v_1|^2 + \ldots + |v_n|^2 = 1$. Suppose that we have constructed $U_1, \ldots, U_{k-1} \in SU(2, K)$
such that

$$\Delta_{k-1}(U_{k-1}) \cdot \ldots \cdot \Delta_1(U_1) e_1 = (v_1, \ldots, v_{k-1}, w_k, 0, \ldots, 0)^T$$
$$=: w_k,$$

with $w_k \in R$. We first show that $1 - \left(\frac{|v_k|}{w_k}\right)^2$ is a sum of squares,
hence a square, in R. Note that by construction

$$w_k^2 = 1 - |v_1|^2 - \ldots - |v_{k-1}|^2 = |v_k|^2 + \ldots + |v_n|^2,$$

since $w_k^* w_k = 1$. Therefore there exists $w_{k+1} \in R$ with

$$w_k^2 - |v_k|^2 = |v_k|^2 + \ldots + |v_n|^2 - |v_k|^2$$
$$= |v_{k+1}|^2 + \ldots + |v_n|^2 = w_{k+1}^2.$$

We conclude that

$$U_k := \frac{1}{w_k} \begin{pmatrix} v_k & -w_{k+1} \\ w_{k+1} & \bar{v}_k \end{pmatrix} \in SU(2, K)$$

and

$$\Delta_k(U_k) w_k = (v_1, \ldots, v_k, w_{k+1}, 0 \ldots, 0)^T.$$

For the last step $k = n - 1$, we make a different choice:

$$U_{n-1} := \frac{1}{w_{n-1}} \begin{pmatrix} v_{n-1} & -\bar{v}_n \\ v_n & \bar{v}_{n-1} \end{pmatrix} \in SU(2, K),$$

because $v_{n-1} \bar{v}_{n-1} + v_n \bar{v}_n = 1 - |v_1|^2 - \ldots - |v_{n-2}|^2 = w_{n-1}^2$. Clearly,

$$\Delta_{n-1}(U_{n-1}) w_{n-1} = (v_1, \ldots, v_n)^T = v.$$

Putting everything together, we arrive at

$$\Delta_{n-1}(U_{n-1}) \cdot \ldots \cdot \Delta_1(U_1)e_1 = v.$$

This proves the lemma. ∎

Proof of (1.20). By definition $S \subseteq SU(n,K)$. The other inclusion is by induction on n. This is clearly true for $n = 2$. Let $U \in SU(n,K)$, and let $v_i := Ue_i$ for $i \in \{1,\ldots,n\}$. Then $v_1,\ldots,v_n$ is an orthonormal basis of K^n. By (1.21) there is $U_1 \in S$ with $U_1e_n = v_n$. Then $U_2 := U_1^{-1}U$ fixes e_n and also $V := (Ke_n)^\perp$, the subspace of K^n orthogonal to e_n. Since V is naturally isomorphic to K^{n-1}, we have $U_2 \in SU(n-1,K)$. By induction hypothesis we conclude that $U_2 \in \left\langle \bigcup_{k=1}^{n-2} \Delta_k(SU(2,K)) \right\rangle$. Hence $U = U_1U_2 \in S$. ∎

Finally, we determine the centralizer of the set of positive definite symmetric matrices of determinant 1.

(1.22) $\mathcal{C}_{GL(n,K)}(SL(n,R) \cap \mathcal{H}(n,K)) = K^*I_n$, *the center of* $GL(n,K)$.

Proof. It is trivial that

$$K^*I_n \subseteq C := \mathcal{C}_{GL(n,K)}(SL(n,R) \cap \mathcal{H}(n,K)).$$

From (1.18) it's easy to see that $C \subseteq \{\text{diag}(\alpha_1,\ldots,\alpha_n); \alpha_k \in K^*\}$. Let's look at the case $n = 2$ first. We have

$$\begin{pmatrix} \alpha_1 & 0 \\ 0 & \alpha_2 \end{pmatrix} \begin{pmatrix} 5 & 2 \\ 2 & 1 \end{pmatrix} \begin{pmatrix} \alpha_1^{-1} & 0 \\ 0 & \alpha_2^{-1} \end{pmatrix} = \begin{pmatrix} 5 & 2\alpha_1\alpha_2^{-1} \\ 2\alpha_2\alpha_1^{-1} & 1 \end{pmatrix}.$$

Therefore $\text{diag}(\alpha_1,\alpha_2) \in C$ implies $\alpha_1 = \alpha_2$, and $C = K^*I_2$.

For the general case we use the embedding $\Delta_k : GL(2,K) \to GL(n,K)$, where $k \in \{1,\ldots,n-1\}$, from above. In fact, a direct application of (1.19) and the just handled case $n = 2$, show that for $\text{diag}(\alpha_1,\ldots,\alpha_n) \in C$ we must have $\alpha_k = \alpha_{k+1}$. This forces $C \subseteq K^*I_n$. The other inclusion has been seen. ∎

H. Formal Power Series

Let R be a commutative ring with 1, and unit group $E := E(R)$. The *Jacobson radical* $J(R)$ of R is the intersection of all maximal ideals in R. We'll have use for the following characterization of $J(R)$ [7; 1.9, p. 6] and [31; Exercise 7.3, p. 205]:

(1.23) *An element $x \in R$ is in $J(R)$ if and only if $1 + xa \in E$ for all $a \in R$. In particular, $1 + I$ is a subgroup of E for every ideal I contained in $J(R)$.* ∎

Let K be a field, and $R := K[[t]]$ the ring of formal power series over K. The following is well-known, see for instance [31; 7.11, p. 195].

(1.24) *R is a local domain with $J(R) = tR$.* ∎

Occasionally we write a formal power series f as

$$f(t) = a_0 + a_1 t + a_2 t^2 + \ldots + a_n t^n + O(t^{n+1}).$$

This means the beginning of f is as stated +terms of order $> n$. Notice that $a_0, \ldots, a_n$ are uniquely determined by f.

(1.25) *Let $x \in tR$, then $1 - x$ is a square in R, and*

$$\sqrt{1-x} := 1 - \frac{1}{2}x - \frac{1}{8}x^2 - \frac{1}{16}x^3 + O\left(x^4\right)$$

is one square root.

Proof. $1 - x$ is a square by Hensel's Lemma [31; Thm. 7.3, p. 185]. The expression for a square root up to higher order terms is easily verified directly. In fact, it coincides with the Taylor expansion over the reals. ∎

As a direct corollary we get for the field of reals $\mathbf{R}$.

(1.26) *The field $\mathbf{R}((t))$ of formal Laurent series over the reals is a pythagorean field, with set of squares*

$$\left\{ \sum_{k=u}^{\infty} a_k t^k;\ u \in 2\mathbf{Z},\ a_u > 0 \right\}.$$

∎

2. Left Loops and Transversals

A set L with a binary operation $L \times L \to L$; $(a,b) \mapsto ab$ is called a *magma*. L will be called a *groupoid* if L contains an *identity* 1, i.e., $\forall a \in L : a1 = 1a = a$. Such an element is necessarily unique. As for groups, we will write $S^{\#} := S \setminus \{1\}$ for every subset S of a groupoid.

Remark. The name "groupoid" occurs in BRUCK's [20] and denotes what is now called a magma. Later, beginning with the 1970-edition of his book "Algebra", BOURBAKI used the word *magma*.

Let L be a magma, then for each $a \in L$, we have a map

$$\lambda_a : L \to L; \; x \mapsto ax,$$

the *left translation*. This gives rise to a map $\lambda : L \to L^L; a \mapsto \lambda_a$. Conversely, given a map $\lambda : L \to L^L; a \mapsto \lambda_a$, one can define a binary operation on L by $ab := \lambda_a(b)$. The existence of an identity 1 is equivalent to

$$\lambda_1 = 1 \quad \text{and} \quad \forall a \in L : \lambda_a(1) = a.$$

In the presence of the second, the first of these conditions can be achieved easily.

(2.1) *Let L be a set with an element $1 \in L$, and let $\mu : L \to L^L; a \mapsto \mu_a$ be a map such that $\mu_a(1) = a$ for all $a \in L$. If μ_1 is bijective, then 1 is an identity of the operation $a \cdot b := \mu_a \mu_1^{-1}(b)$, i.e., $(L, \cdot)$ is a groupoid.* ∎

The *right translations* ϱ_a of a groupoid (or magma) are introduced dually, i.e., $\varrho_a(x) = xa$. With obvious modifications the above statements (and much of what follows) apply to right translations as well.

Let a be an element of a groupoid L. An element $b \in L$ is called a *left (right) inverse* of a if $ba = 1$ ($ab = 1$). If b is the uniquely determined left and right inverse of a, then we speak of the *inverse* of a and write $b = a^{-1}$.

For $a \in L, n \in \mathbf{N}$, we put recursively

$$a^0 := 1, \; a^n := a(a^{n-1}) \quad \text{and} \quad a^{-n} := (a^{-1})^n \quad \text{if } a^{-1} \text{ exists.}$$

This clearly implies $\lambda_a^n(1) = a^n$.

Since groupoids are in general non-associative, we'll have to use lots of parentheses. To save a few, we shall adopt the following *dot-convention*:

$$a \cdot bc = a(bc), \quad ab \cdot c = (ab)c, \quad \text{and} \quad a^k b = (a^k)b, \quad ba^k = b(a^k)$$

for $a, b, c \in L, k \in \mathbf{Z}$. E.g., the associative law reads $a \cdot bc = ab \cdot c$.

An element a from a groupoid L is called

left (right) alternative if $a \cdot ab = a^2 b$ $(ba \cdot a = ba^2)$ $\forall b \in L$;

left power alternative if a and λ_a each have an inverse and
$\lambda_a^k = \lambda_{a^k}, \forall k \in \mathbf{Z}$.

It is said to have the

left (right) inverse property if there exists $a' \in L$ with
$a' \cdot ab = b$ $(ba \cdot a' = b)$ $\forall b \in L$.

If every element of a groupoid L has one of the above properties, then the corresponding phrase will also be used for L. Moreover, we say that L

is a *left (right) loop* if there is a unique solution $x \in L$ of the
equation $ax = b$ $(xa = b)$ for all $a, b \in L$;

is a *loop* if it is a left and a right loop (recall that $1 \in L$);

is *Bol* if $a(b \cdot ac) = (a \cdot ba)c$ for all $a, b, c \in L$;

satisfies the *automorphic inverse property* if all $a, b \in L$ have
inverses, and $(ab)^{-1} = a^{-1}b^{-1}$;

is a *K-loop* if it is a Bol loop and satisfies the automorphic inverse
property.

Remark. Some authors use the phrase "right loop" for left loops, and vice versa (e.g., [81]). Moreover, some authors only assume the presence of a right identity ε in a left loop, i.e., $a\varepsilon = a$ for all a in L. The literature is not consistent here.

For completeness, we record a few observations. They are all easy consequences of the definitions.

(2.2) Let L be a groupoid and $a \in L$.

(1) *If every element in L has a (unique!) inverse, then* $(a^{-1})^{-1} = a$.

(2) *a is left alternative if and only if $\lambda_a^2 = \lambda_{a^2}$.*

(3) *L is a left loop if and only if λ_b is bijective for every $b \in L$.* ∎

We shall use this frequently without specific reference.

A. LEFT LOOPS AND THE LEFT MULTIPLICATION GROUP

In a left loop L, all the maps λ_a are bijective. Therefore, $\lambda(L) := \{\lambda_a; a \in L\}$ generates a group

$$\mathcal{M}_\ell = \mathcal{M}_\ell(L) := \langle \lambda_a; a \in L \rangle$$

called the *left multiplication group* of L. In fact, $\mathcal{M}_\ell$ is a subgroup of $\mathcal{S}_L$, the symmetric group on the set L. In this setting, it makes sense to define for all $a, b \in L$ the *precession map*

$$\delta_{a,b} := \lambda_{ab}^{-1} \lambda_a \lambda_b.$$

Obviously, these mappings are characterized by the property

$$a \cdot bx = ab \cdot \delta_{a,b}(x) \quad \text{for all } x \in L.$$

Furthermore, let

$$\mathcal{D} = \mathcal{D}(L) := \langle \delta_{a,b}; a, b \in L \rangle$$

be the subgroup of $\mathcal{M}_\ell$ generated by all precession maps $\delta_{a,b}$, $a, b \in L$. The set $\mathcal{D}$ is called the *left inner mapping group*. In [81] the name *structure group*[1] has been used. The elements of $\mathcal{D}$ are called *left inner mappings*. Notice that left inner mappings fix 1. In accordance with a definition from §1, we shall say that a subset S of $\mathcal{D}$ is *fixed point free* if it acts fixed point free on $L^\#$.

For completeness and for easier referencing we record

---

[1] It does not say very much about the structure of L, e.g., every group has $\mathcal{D} = \{1\}$.

(2.3) *For a left loop L the following are equivalent*

    (I)   *L is associative;*

   (II)   *L is a group;*

  (III) *The map $\lambda : L \to \mathcal{M}_\ell$ is an isomorphism;*

  (IV) *$\mathcal{M}_\ell = \lambda(L)$;*

   (V) *$\mathcal{D} = \{\mathbf{1}\}$.*

Proof. (I) $\Longrightarrow$ (II): For every $a \in L$ there exists $a' \in L$ such that $aa' = 1$. Therefore, L is a group.

(II) $\Longrightarrow$ (III): For $a, b, c \in L$ we have $a \cdot bc = ab \cdot c$, hence $\lambda_a \lambda_b = \lambda_{ab}$, and λ is monomorphism. Thus $\lambda(L)$ is a group, hence $\lambda(L) = \mathcal{M}_\ell$.

"(III) $\Longrightarrow$ (IV)" and "(V) $\Longrightarrow$ (I)" are trivial.

(IV) $\Longrightarrow$ (V): Let $\alpha \in \mathcal{D}$, then $\alpha(1) = 1$. By assumption there exists $a \in L$ with $\alpha = \lambda_a$. This implies $a = 1$, and $\alpha = \mathbf{1}$.   ■

We put down some basic properties of automorphisms. As for groups, the reader convinces herself that for every automorphism α of the groupoid L we have $\alpha(1) = 1$ (always), and $\alpha(a^{-1}) = \alpha(a)^{-1}$ if every $a \in L$ has a (unique) inverse.[2] These observations will be used freely throughout this work.

(2.4) *Let α be a permutation of the groupoid L.*

(1) *The following are equivalent*

    (I)   *$\alpha \in \operatorname{Aut} L$;*

   (II) *$\alpha(1) = 1$ and $\forall a \in L : \alpha \lambda_a \alpha^{-1} \in \lambda(L)$;*

  (III) *$\forall a \in L : \alpha \lambda_a \alpha^{-1} = \lambda_{\alpha(a)}$.*

(2) *If L is a left loop, $\alpha \in \operatorname{Aut} L$, and $a, b \in L$, then $\alpha \delta_{a,b} \alpha^{-1} = \delta_{\alpha(a),\alpha(b)}$.*

Proof. (1) (I) $\Longrightarrow$ (II): $\alpha(1) = 1$ is true by a previous remark. Furthermore,

$$\alpha \lambda_a \alpha^{-1}(x) = \alpha(a\alpha^{-1}(x)) = \alpha(a)x = \lambda_{\alpha(a)}(x) \quad \text{for all } x \in L.$$

---

[2] The existence of inverses can be relaxed to the existence of unique right inverses if L is a left loop, and the like. We have not tried to be most general.

Therefore $\alpha\lambda_a\alpha^{-1} \in \lambda(L)$.

(II) $\implies$ (III): $\alpha\lambda_a\alpha^{-1} = \lambda_c$ for some $c \in L$. Applying both sides to 1 and using $\alpha^{-1}(1) = 1$ yields $\alpha(a) = c$.

(III) $\implies$ (I): If $a, b \in L$, then $\alpha(ab) = \alpha\lambda_a(b) = \alpha\lambda_a\alpha^{-1}\alpha(b) = \lambda_{\alpha(a)}\alpha(b) = \alpha(a)\alpha(b)$.

(2) is a direct consequence of (1) and the definition of $\delta_{a,b}$. $\blacksquare$

Part (1) of the preceding theorem for loops is due to ALBERT [2; §9 and Thm. 9]. The following, which is [2; Thm. 8], is an easy corollary.

(2.5) *The map* $\mathrm{Aut}\, L \to \mathrm{Aut}\, \mathcal{M}_\ell(L);\ \alpha \mapsto \hat{\alpha}$ *is a monomorphism.*
$\blacksquare$

Remarks. 1. ALBERT gives a generalization of (2.4.1) for anti-automorphisms in [2; Thm. 10]. We will not dwell on this.

2. (2.4.1) can also be generalized to arbitrary homomorphisms. This is left to the reader.

By duality, for right loops L we'd have to consider

$$\mathcal{M}_r := \langle \varrho_a;\ a \in L \rangle,$$

the *right multiplication group,* and would get analogous results. If L is a loop one can also study the (full) *multiplication group* $\mathcal{M} := \langle \mathcal{M}_\ell \cup \mathcal{M}_r \rangle$. This has been done for instance by BRUCK in [20].

B. TRANSVERSALS AND SECTIONS

Let G be a group with a subgroup Ω. A set L of representatives of the left cosets of Ω in G with $1 \in L$, will be called a *transversal* of G/Ω. More precisely, we require that

$$L \subseteq G \quad \text{such that} \quad \forall g \in G: |L \cap g\Omega| = 1 \quad \text{and} \quad 1 \in L.$$

By abuse of language, we'll simply refer to L as a "transversal of the *coset space* G/Ω". This is meant to express that G is a group with a subgroup Ω and a transversal L.

By the axiom of choice, transversals always exist. We have

(2.6) *Let L be a left loop.*

(1) $\mathcal{D}(L) = \{\alpha \in \mathcal{M}_\ell; \alpha(1) = 1\}.$

(2) $\lambda(L)$ *is a transversal of* $\mathcal{M}_\ell / \mathcal{D}.$

Proof. (1) Let $S := \{\alpha \in \mathcal{M}_\ell; \alpha(1) = 1\}$. It is clear that $\mathcal{D} \subseteq S$.

Let $K := \{\alpha \in \mathcal{M}_\ell; \alpha \in \lambda_{\alpha(1)} \mathcal{D}\}$. We shall prove that $K = \mathcal{M}_\ell$. This will imply for $\alpha \in S$ that $\alpha \in \lambda_{\alpha(1)} \mathcal{D} = \mathcal{D}$, hence $\mathcal{D} = S$.

K is not empty, since $1 \in K$. For $b \in L, \alpha \in K$, there exists $\beta \in \mathcal{D}$ such that

$$\lambda_b \alpha = \lambda_b \lambda_{\alpha(1)} \beta = \lambda_{b\alpha(1)} \delta_{b,\alpha(1)} \beta \in K,$$

and for $c \in L$ with $\alpha(1) = bc$ (i.e., $c = \lambda_b^{-1} \alpha(1)$)

$$\lambda_b^{-1} \alpha = \lambda_b^{-1} \lambda_{\alpha(1)} \beta = \lambda_c \lambda_c^{-1} \lambda_b^{-1} \lambda_{bc} \beta = \lambda_c \delta_{b,c}^{-1} \beta \in K.$$

Now, every element of $\mathcal{M}_\ell$ is a finite product of λ_b's and λ_b^{-1}'s, thus we can conclude that $\mathcal{M}_\ell K \subseteq K$. This implies $\mathcal{M}_\ell = K$, which is all we needed to show.

(2) By (1) we have that $\lambda_a \mathcal{D}$, $a \in L$, contains exactly the elements α of $\mathcal{M}_\ell$ with $\alpha(1) = a$. Therefore, $\lambda_a \mathcal{D} = \lambda_b \mathcal{D} \iff a = b$. On the other hand, given $\alpha \in \mathcal{M}_\ell$, we have $\alpha \in \lambda_{\alpha(1)} \mathcal{D}$. Hence the result. ∎

Remarks. 1. Statement (1) occurs in a slightly different context as [20; IV Lemma 1.2 p. 61]. The proof is adapted to our situation (see also [67; Prop. 1.1]).

2. If L is a transversal of G/Ω, then $G = L\Omega$, where each $g \in G$ has a unique decomposition $g = a\omega$, $a \in L$, $\omega \in \Omega$. Therefore, in [81] $G = L\Omega$ has been called a *direct decomposition*.

As a converse of (2.6), we'll now show how a transversal always has the structure of a left loop. In the future, we'll tacitly assume that transversals are endowed with this structure.

(2.7) Theorem. *Let L be a transversal of the coset space G/Ω. We have*

(1) *For $a, b \in L$ there are unique $a \circ b \in L$ and $d_{a,b} \in \Omega$ such that $ab = (a \circ b)d_{a,b}.$*

(2) $(L, \circ)$ *is a left loop.*

(3) *If G' is a subgroup of G containing L, i.e., $L \subseteq G'$, and $\Omega' := G' \cap \Omega$, then L is a transversal of G'/Ω' and the multiplications obtained from G/Ω and G'/Ω' coincide.*

(4) *The following are equivalent*

    (I)   *L is a loop;*

   (II)   *For every $a, b \in L$ we have $|La \cap b\Omega| = 1$;*

  (III)   *For every $g \in G$, L is a transversal of $g\Omega g^{-1}$;*

  (IV)   *For every $g \in G$, gLg^{-1} is a transversal of Ω;*

   (V)   *For every $a \in L$, $a^{-1}La$ is a transversal of Ω.*

(5) *Two elements $a, b \in L$ commute, i.e., $a \circ b = b \circ a$ if and only if $[a, b] = a^{-1}b^{-1}ab \in \Omega$.*

(6) *Let L' be a transversal of another coset space G'/Ω'. If $\alpha : G \to G'$ is a homomorphism such that $\alpha(\Omega) \subseteq \Omega'$ and $\alpha(L) \subseteq L'$, then $\alpha|_L : L \to L'$ is a homomorphism of loops, and $\alpha(d_{a,b}) = d_{\alpha(a),\alpha(b)}$.[3] In particular, for $\omega \in \Omega$ we have $\hat{\omega} \in \mathrm{Aut}\, L$ if and only if $\omega L \omega^{-1} = L$.*

Proof. (1) is clear from the definition.

(2) It is easy to see that 1 is an identity with respect to "$\circ$".

Let $a, b \in L$, and consider the equation $a \circ x = b$. We can decompose $a^{-1}b = x_0\omega$, with $x_0 \in L$, $\omega \in \Omega$, and obtain $(a \circ x_0)d_{a,x_0} = ax_0 = bw^{-1}$. Hence $a \circ x_0 = b$, by uniqueness of the decomposition. Let $x_1 \in L$ be another solution, then $ax_1 d_{a,x_1}^{-1} = b = ax_0\omega$, and $x_1 d_{a,x_1}^{-1} = x_0\omega$. Uniqueness of the decomposition shows $x_0 = x_1$.

(3) The only assertion, which is not completely trivial is $L \cap g\Omega' \neq \emptyset$ for every $g \in G'$. Now for $g \in G'$ there exists $\omega \in \Omega$ with $g\omega \in L$. This implies $\omega \in g^{-1}L \subseteq G'$, hence $\omega \in G' \cap \Omega = \Omega'$.

The last statement is clear, too, since $d_{a,b} \in \Omega'$.

(4) (I) $\Longrightarrow$ (II): For $a, b \in L$ there exists a unique $x \in L$ such that $x \circ a = b$. Hence $La \cap b\Omega = \{xa\}$.

---

[3] For briefness we do not distinguish the d's formed in G and G' in our notation.

(II) $\Longrightarrow$ (III): Let $g, h \in G$. There exist $a, b \in L$, $\omega \in \Omega$ such that $g = a\omega$ and $b \in hg\Omega$. Now

$$|L \cap hg\Omega g^{-1}| = |Lg \cap hg\Omega| = |Lg\omega^{-1} \cap b\Omega\omega^{-1}| = |La \cap b\Omega| = 1.$$

Since $1 \in L$ anyway, we're done.

(III) $\Longrightarrow$ (IV): Apply the inner automorphism $\widehat{g^{-1}} : x \mapsto g^{-1}xg$ to see that $g^{-1}Lg$ is a transversal of Ω.

"(IV) $\Longrightarrow$ (V)" is trivial.

(V) $\Longrightarrow$ (I): Let $a, b \in L$. By assumption, $|a^{-1}La \cap a^{-1}b\Omega| = 1$, i.e., there exist exactly one $x \in L$, and exactly one $d \in \Omega$, such that $a^{-1}xa = a^{-1}bd$. Thus x is the unique solution of the equation $x \circ a = b$.

(5) Assume first that a, b commute, then

$$abd_{a,b}^{-1} = a \circ b = b \circ a = bad_{b,a}^{-1} \implies [a, b] \in \Omega.$$

Conversely, $[a, b] \in \Omega$ implies $ba \in ab\Omega$, hence

$$bad_{b,a}^{-1}, \; abd_{a,b}^{-1} \in L \cap ab\Omega \implies bad_{b,a}^{-1} = abd_{a,b}^{-1},$$

and a, b commute.

(6) Let $a, b \in L$. We decompose $\alpha(a)\alpha(b)$ according to (1) in two ways

$$\alpha(a)\alpha(b) = \alpha(ab) = \alpha((a \circ b)d_{a,b}) = \alpha(a \circ b)\alpha(d_{a,b})$$
$$= (\alpha(a) \circ' \alpha(b))d'_{\alpha(a),\alpha(b)},$$

where $\circ'$ and d' refer to L'. The uniqueness of the decomposition and the assumptions imply $\alpha(a \circ b) = \alpha(a) \circ' \alpha(b)$ and $\alpha(d_{a,b}) = d'_{\alpha(a),\alpha(b)}$. The last statement is now obvious.   ■

The symbol $d_{a,b}$ will be used with the meaning of the theorem whenever it makes sense in a given context. We'll not comment on this any further.

Given a transversal L of G/Ω, one can define a map

$$\sigma_L : G/\Omega \to G; \; A \mapsto \sigma_L(A) \quad \text{by} \quad A \cap L = \{\sigma_L(A)\}.$$

This makes sense because $A \cap L$ has exactly one element for every left coset A. Notice that $A = \sigma_L(A)\Omega$, so σ_L "chooses" a representative for each coset of Ω in G. Now

$$A * B := \sigma_L(A)B, \quad A, B \in G/\Omega$$

defines a left loop multiplication on G/Ω. In fact, it is straightforward to verify that

$$\sigma_L : (G/\Omega, *) \to (L, \circ)$$

is an isomorphism. So the preceding remark gives only a different description of the construction in (2.7). Note also that using σ_L it's particularly easy to write up a proof of (2.7.2).

Conversely, for a group G with a subgroup Ω, a map $\sigma : G/\Omega \to G$ is called a *section* if $\sigma(\Omega) = 1$ and $\pi\sigma = \mathbf{1}$, where $\pi : G \to G/\Omega$; $g \mapsto g\Omega$ is the quotient map. Clearly, $\sigma(G/\Omega)$ is a transversal of G/Ω. The corresponding section $\sigma_{\sigma(G/\Omega)}$ as above is σ again. Note that σ_L is a section, and $\sigma_L(G/\Omega) = L$. Thus there is no essential difference in using transversals or sections. Some authors call the set $\sigma(G/\Omega)$ a section.

Abusing language again, we'll briefly say: "σ is a section of the coset space G/Ω".

In the sequel, we'll switch freely between transversals and their corresponding sections, somewhat biased to transversals, though.

Finally, we remark that sections have also been called "cross sections" in the literature (e.g., in [46]).

One advantage of the use of sections is that they allow to express topological conditions more easily, and sometimes more naturally (see e.g., [46] or [88]).

We'll now establish connections between various objects inside a group and objects coming with the left loop structure on a transversal in that group.

So let L be a transversal of the coset space G/Ω, with corresponding section $\sigma : G/\Omega \to G$, i.e., $\sigma(G) = L$. We can use σ

to induce a permutation representation of G on L, by putting

$$\theta : \begin{cases} G \to S_L \\ g \mapsto \theta_g : \begin{cases} L \to L \\ a \mapsto \sigma(ga\Omega). \end{cases} \end{cases} \qquad \begin{array}{ccc} G/\Omega & \xrightarrow{g} & G/\Omega \\ \downarrow \sigma & & \downarrow \sigma \\ L & \xrightarrow{\theta_g} & L \end{array}$$

The construction of θ is illustrated by the diagram. Notice that $L \to G/\Omega$; $a \mapsto a\Omega$ inverts σ, thus $\sigma : G/\Omega \to L$ is a bijection.

(2.8) Theorem. *Using the notation just introduced, we have*

(1) θ *is a homomorphism with kernel* $N = \bigcap_{h \in G} h\Omega h^{-1}$. *Thus, N is the core of Ω in G. In other words, θ is a permutation representation of G on L, which is equivalent to the natural permutation representation of G on G/Ω via* $(\mathbf{1}, \sigma)$.

(2) $\theta|_L = \lambda$, *i.e., for $a, b \in L$ we have $\theta_a(b) = a \circ b$. In particular, $\theta|_L$ is injective.*

(3) $\theta_{d_{a,b}} = \delta_{a,b}$, *for all $a, b \in L$, where $d_{a,b} = (a \circ b)^{-1} ab \in \Omega$ (see (2.7)).*

(4) *If the set L generates G as a group, then*

$$\theta(G) = \mathcal{M}_\ell(L) \quad \text{and} \quad \theta(\Omega) = \mathcal{D}(L), \quad \text{hence} \quad \mathcal{D}(L) \cong \Omega/N.$$

(5) *If Ω is a normal subgroup of G, then $\sigma : G/\Omega \to L$ is an isomorphism, and L is a group. Conversely, if the set L generates G as a group, and $(L, \circ)$ is a group, then Ω is a normal subgroup of G.*

(6) *For $\omega \in \Omega$ if $\omega L \omega^{-1} \subseteq L$, then $\theta_\omega = \hat{\omega} \in \mathrm{Aut}\, L$. If $\omega L \omega^{-1} \subseteq L$ for all $\omega \in \Omega$, then*

$$\mathcal{D}(L) \subseteq \theta(\Omega) = \hat{\Omega} \subseteq \mathrm{Aut}\, L.$$

Note that $\theta(\Omega) \cong \Omega/N$, and $N = \mathcal{C}_\Omega(L)$.

(7) *If $\sigma' : G/\Omega \to G$ is another section and $L' := \sigma'(G/\Omega)$, then the following are equivalent*

    (I)   *the multiplications induced on G/Ω by σ and σ' coincide;*

    (II) $\sigma' \sigma^{-1} : L \to L'$ *is an isomorphism of left loops;*

    (III) *there exists a map $\Xi : L \to L'$ such that $aN = \Xi(a)N$ for all $a \in L$;*

(IV) $\forall A \in G/\Omega : \sigma'(A) \in \sigma(A)N$.

In the above situation, we necessarily have $\Xi = \sigma'\sigma^{-1}$.

Proof. (1) Let $g, h \in G, a \in L$, then

$$\theta_g\theta_h(a) = \sigma\big(g\sigma(ha\Omega)\Omega\big) = \sigma(gha\Omega) = \theta_{gh}(a).$$

Thus θ is a homomorphism. In particular, θ_g is invertible.

$g \in G$ is in the kernel of θ if and only if

$$\forall a \in L : \theta_g(a) = a \iff \forall a \in L : ga\Omega = a\Omega$$
$$\iff \forall h \in G : gh \in h\Omega \iff \forall h \in G : h^{-1}gh \in \Omega$$
$$\iff g \in \bigcap_{h \in G} h\Omega h^{-1},$$

since L is a transversal of G. So N is indeed the core of Ω in G.

Finally, it is direct to see that $\sigma : G/\Omega \to L$ induces the given equivalence (see also the diagram just before the statement of the theorem).

(2) For $a, b \in L$ we have $\theta_a(b) = \sigma(ab\Omega) = a \circ b$, by definition.

(3) $\theta_{d_{a,b}} = \theta_{(a \circ b)^{-1}ab} = \theta_{(a \circ b)}^{-1}\theta_a\theta_b = \lambda_{(a \circ b)}^{-1}\lambda_a\lambda_b = \delta_{a,b}$, by (1) and (2).

(4) From (2) follows $\theta(L) = \lambda(L) \subseteq M_\ell$. Thus a generating set of G is mapped onto a generating set of M_ℓ. This implies $\theta(G) = M_\ell$.

For all $g \in G$ we have

$$\theta_g(1) = 1 \iff \sigma(g\Omega) = 1 = \sigma(\Omega) \iff g \in \Omega.$$

Together with (2.6) this implies $\theta(\Omega) = \mathcal{D}(L)$. The last assertion is now clear.

(5) For $a, b \in L$ we have $\sigma(a\Omega b\Omega) = \sigma(ab\Omega) = a \circ b$. Since L is a transversal, σ is a homomorphism. It is bijective by definition.

If L is a group, then $\mathcal{D}(L) = \{1\}$, (2.3). By assumption we can apply (4), to conclude that $\Omega = N$. Hence Ω is normal.

(6) The first statement follows from (2.7.6), since for all $a \in L$ we have

$$\theta_\omega(a) = \sigma(\omega a \omega^{-1}\Omega) = \omega a \omega^{-1} = \hat{\omega}(a).$$

Together with (3) we obtain $\mathcal{D} \subseteq \theta(\Omega) \subseteq \operatorname{Aut} L$. The remaining claims are trivial.

(7) (I) $\Longrightarrow$ (II): Since both $\sigma : G/\Omega \to L$ and $\sigma' : G/\Omega \to L'$ are isomorphisms, then so is $\sigma'\sigma^{-1}$.

(II) $\Longrightarrow$ (III): For $a, b \in L$ we have

$$\sigma'(ab\Omega) = \sigma'\sigma^{-1}(a \circ b) = \sigma'\sigma^{-1}(a) \circ \sigma'\sigma^{-1}(b)$$
$$= \sigma'\left(\sigma'\sigma^{-1}(a)\sigma'(b\Omega)\Omega\right).$$

σ' is injective, so we can conclude

$$ab\Omega = \sigma'\sigma^{-1}(a)b\Omega, \quad \text{hence} \quad a \in \sigma'\sigma^{-1}(a)b\Omega b^{-1}.$$

This is true for arbitrary $b \in L$, and since L is a transversal, it is in fact true for every $b \in G$. Thus $\Xi := \sigma'\sigma^{-1}$ qualifies.

(III) $\Longrightarrow$ (IV): For $A \in G/\Omega$, we have

$$\Xi\sigma(A) \in \sigma(A)N \subseteq \sigma(A)\Omega = A.$$

Therefore, $\Xi\sigma(A) = \sigma'(A)$, and $\sigma'(A) \in \Xi\sigma(A)N = \sigma(A)N$.

(IV) $\Longrightarrow$ (I): For $A, B \in G/\Omega$ let $n \in N$ be such that $\sigma'(A) = \sigma(A)n$. Then for any $b \in B$

$$\sigma'(A)B = \sigma(A)nb\Omega = \sigma(A)b(b^{-1}nb)\Omega = \sigma(A)b\Omega = \sigma(A)B.$$

Hence the two multiplications on G/Ω coincide.

For the final remark we observe that $\Xi\sigma(a\Omega) \in L'$, hence $\Xi\sigma(a\Omega) = \sigma'(a\Omega)$ for all $a \in L$. Injectivity implies $\Xi\sigma = \sigma'$, and $\Xi = \sigma'\sigma^{-1}$. ∎

If the two transversals L, L', or the two sections σ, σ' satisfy the conditions in (7) of the preceding theorem, they are called *congruous*. From the point of view of loop theory, it makes no essential difference to pass from a transversal (or section) to a congruous one.

If $N = \{1\}$, then "congruous" is the same as "equal". Similarly, there is no loss of generality to considering G/N instead.

θ is called the *natural permutation representation* of G on L. By the permutation group (G, L) we'll always refer to the natural permutation representation.

Remarks. 1. In general, the image of θ is strictly bigger than $\mathcal{M}_\ell(L)$.

2. Parts (1), (2) and (5) of the last theorem are taken from [8; 1.1, 2.5], (4) and (7) are generalizations of [8; (1.5), 2.3]). We have modernized and simplified exposition and proof. Part (6) generalizes [81; (3.7)].

3. (12.3.1) provides examples of left loops with (2.8.6), which are not loops.

The left loops which satisfy (2.8.6) are so important that they are given a name:
A left loop L is called A_ℓ, or A_ℓ-*left-loop* if $\mathcal{D}(L) \subseteq \text{Aut}(L)$. This is equivalent to the (formally weaker) requirement that $\delta_{a,b}$ be an automorphism of L for every $a, b \in L$. The "ℓ" in the index refers to the fact that $\mathcal{D}$ was defined to be the *left* inner mapping group of L. A loop with A_ℓ will be called an A_ℓ-*loop*. They have also been called *weak K-loops* or *WK-loops*.

If the right inner mappings (defined analogously using right translations), and all maps of the form $\lambda_a \varrho_a^{-1}, a \in L$, are also automorphisms, then the loop is called an *A-loop*. In [19] the theory of A-loops has been studied.

The construction in the theorem can be seen in a (formally) more general perspective: Let G be a group acting transitively on a set P. If Ω is the point stabilizer of $e \in P$, say, then the map

$$\phi : \begin{cases} G/\Omega \to P \\ \gamma\Omega \mapsto \gamma(e) \end{cases}$$

gives an equivalence $(\mathbf{1}, \phi)$ of the permutation groups $(G, G/\Omega)$ and (G, P), see (1.3.2). With this premise we obtain a corollary to (2.8.1) (which is illustrated by a diagram):

(2.9) Let $\sigma : G/\Omega \to G$ be a section with corresponding transversal $L := \sigma(G/\Omega)$. Then $(1_G, \sigma\phi^{-1})$ is an equivalence from (G, P) to (G, L). ∎

$$
\begin{array}{ccc}
P & \xrightarrow{\gamma} & P \\
\uparrow \phi & \gamma & \uparrow \phi \\
G/\Omega & \xrightarrow{\gamma} & G/\Omega \\
\downarrow \sigma & & \downarrow \sigma \\
L & \longrightarrow & L
\end{array}
$$

The left loop structure of L can be carried over to P via the (bijective) map $\phi\sigma^{-1}|_L$. Here's another way to describe the resulting multiplication on P.

(2.10) Theorem. Let (G, P) be a permutation group. For fixed $e \in P$, let Ω be the stabilizer of e in G and let $\mu : P \to G; x \mapsto \mu_x$ be a map such that $\mu_x(e) = x$ for all $x \in P$.

(1) (G, P) is transitive.

(2) For every section $\sigma : G/\Omega \to G$ we have $\phi\sigma^{-1}(\alpha) = \alpha(e)$ for all $\alpha \in \sigma(G/\Omega)$.

(3) $\mu(P)$ is a set of representatives of the left cosets of Ω in G, and $L := \mu(P)\mu_e^{-1}$ is a transversal of the coset space G/Ω, i.e., L is a left loop.

(4) If $\mu_e = 1$, let $x \bullet y := \mu_x(y)$ for all $x, y \in P$, then the map $\phi\sigma^{-1} : L \to (P, \bullet)$ is an isomorphism, where σ is the section corresponding to $L = \mu(P)$. Moreover, $\mu : P \to L$ and $\phi\sigma^{-1}|_L$ are inverse mappings.

Proof. (1) is clear; and (2) follows directly from the definitions.

(3) For $\gamma \in G$, we have $\gamma^{-1}\mu_{\gamma(e)} \in \Omega$, and $\mu(P) \cap \gamma\Omega = \{\mu_{\gamma(e)}\}$. Therefore, $\mu(P)$ is a set of representatives of the left cosets of Ω in G. Likewise $L \cap \gamma\Omega = \{\mu_{\gamma(e)}\mu_e^{-1}\}$. Moreover, $1 \in L$, thus L is a transversal.

(4) We'll use (2) freely. For $x \in P$ we have $\phi\sigma^{-1}\mu_x = \mu_x(e) = x$, thus $\phi\sigma^{-1}\mu = 1$. For $\alpha \in L$, we find $\mu\phi\sigma^{-1}(\alpha) = \mu_{\alpha(e)} = \alpha$, since there is exactly one element in L which maps e to $\alpha(e)$. Therefore $\mu\phi\sigma^{-1} = 1$, as well. This shows the last assertion; in particular, $\phi\sigma^{-1}$ is bijective.

Denoting the multiplication in L by $\circ$, we compute for all $\alpha, \beta \in L$
$$
\phi\sigma^{-1}(\alpha \circ \beta) = \phi\sigma^{-1}(\sigma(\alpha\beta\Omega)) = \phi(\alpha\beta\Omega) = \alpha\beta(e)
$$
$$
\overset{!}{=} \mu_{\alpha(e)}(\beta(e)) = \alpha(e) \bullet \beta(e).
$$
Using (2), this shows that $\phi\sigma^{-1}$ is a homomorphism. ∎

In the situation of (4) in the theorem $(P, \bullet)$ is called the *natural left loop structure* on P.

Now assume for a moment that L is an arbitrary subset of G which contains 1 and acts regularly on P. Then for all $x \in P$, we have a unique $\mu_x \in L$ such that $\mu_x(e) = x$. As in the preceding lemma we put

$$x \bullet y := \mu_x(y) \quad \text{for all } x, y \in P.$$

Then $(P, \bullet)$ is a left loop with identity e. In fact, P is a loop:

(2.11) Theorem. *Let (G, P) be a permutation group. For fixed $e \in P$, let Ω be the stabilizer of e.*

(1) *If L is a subset of G, containing 1, which acts regularly on P, then L is a transversal of G/Ω. Furthermore, L is a loop, and the map $\phi\sigma^{-1} : L \to (P, \bullet)$ is an isomorphism, where σ is the section corresponding to L.*

(2) *If L is a loop transversal of the coset space G/Ω, then L acts regularly on P.*

(3) *If L' is a subset of G, acting regularly on P, then $L := L'\alpha^{-1}$ satisfies the hypothesis of (1) for arbitrary $\alpha \in L'$. In particular, L is a loop transversal.*

Proof. (1) In view of (2.10), we only have to prove that for all $a, b \in P$ the equation $x \bullet a = b$ has a unique solution x in P. Indeed, there exists exactly one $\alpha \in L$ with $\alpha(a) = b$, since L acts regularly. Therefore $\alpha(e)$ is the desired solution.

(2) Let σ be the section corresponding to L. For $a, b \in P$ put $\alpha := \sigma\phi^{-1}(a), \beta := \sigma\phi^{-1}(b) \in L$. By hypothesis there exists a unique $\xi \in L$ with $\xi \circ \alpha = \beta$. From (1) we have $\alpha(e) = a$ and $\beta(e) = b$, therefore we can compute

$$\xi(a) = \xi\alpha(e) = \phi(\xi\alpha\Omega) = \phi\sigma^{-1}(\xi \circ \alpha) = \phi\sigma^{-1}(\beta) = \beta(e) = b.$$

The uniqueness of ξ comes from the uniqueness of the solution in L.

(3) Clearly, $1 \in L$. For $a, b \in P$ there exists $\gamma \in L$ with $\gamma(\alpha^{-1}(a)) = b$. Hence L acts regularly on P. ∎

C. Transassociants and the Quasidirect Product

If L is a left loop with left inner mapping group $\mathcal{D} = \mathcal{D}(L)$, the group $\mathcal{M}_\ell$ can be constructed directly. The procedure is due to Sabinin [105, 106]. His main idea was to put the "obstruction" into the map

$$\chi : \begin{cases} L \times \mathcal{S}_L \to \mathcal{S}_L \\ (a, \alpha) \mapsto \lambda_{\alpha(a)}^{-1} \alpha \lambda_a \alpha^{-1} \end{cases} .$$

Before we present Sabinin's construction, we collect a few simple observations.

(2.12) *Let $a \in L$ and $\alpha \in \mathcal{S}_L$. We have*

(1) $\alpha \in \mathcal{M}_\ell \implies \lambda_{\alpha(a)}^{-1} \alpha \lambda_a \in \mathcal{D}$.

(2) $\chi(L \times \mathcal{M}_\ell) \subseteq \mathcal{M}_\ell$ *and* $\chi(L \times \mathcal{D}) \subseteq \mathcal{D}$.

(3) $\alpha \in \operatorname{Aut} L \iff \forall a \in L : \chi(a, \alpha) = \mathbf{1}$.

Proof. (1) The map $\lambda_{\alpha(a)}^{-1} \alpha \lambda_a$ is clearly in $\mathcal{M}_\ell$, and fixes 1. By (2.6.1) it is an element of $\mathcal{D}$.

(2) is clear from (1); and (3) is just a rewording of (2.4.1).   ∎

In view of (3), χ can also be seen as a measure how much a map $\alpha \in \mathcal{S}_L$ deviates from an automorphism.

Subsets T of $\mathcal{S}_L$ with $\chi(L \times T) \subseteq T$ will be called χ-*invariant*. Following Sabinin (see [88]), we call a χ-invariant subgroup of $\mathcal{S}_L$ which fixes 1 and contains $\mathcal{D}(L)$ a *transassociant* of L. We emphasize that transassociants are groups by definition. By the preceding lemma $\mathcal{D}$, $\operatorname{Aut} L$, and $\mathcal{M}_\ell$ are examples of χ-invariant sets. $\mathcal{D}$ is a transassociant, and $\operatorname{Aut} L$ is a transassociant if and only if $\mathcal{D} \subseteq \operatorname{Aut} L$, i.e., L is A_ℓ.

The main point of these definitions is the construction of the quasidirect product.

(2.13) Theorem. *Let $(L, \cdot)$ be a left loop, and let T be a transassociant of L. Then*

(1) $L \times_Q T$, *the set $L \times T$ with the multiplication*

$$(a, \alpha)(b, \beta) := \big(a \cdot \alpha(b), \delta_{a, \alpha(b)} \chi(b, \alpha) \alpha \beta\big),$$

for all $(a, \alpha), (b, \beta) \in L \times T$, is a group. The inverse of $(a, \alpha) \in L \times_Q T$ is given by

$$(a, \alpha)^{-1} = \left(\alpha^{-1}(a'), \, \alpha^{-1}\chi(\alpha^{-1}(a'), \alpha)^{-1} \delta^{-1}_{a, a'} \right),$$

where $a' = \lambda_a^{-1}(1)$ is the right inverse of a in L.

(2) $L \times_Q T$ acts faithfully and transitively on L by

$$(a, \alpha)(x) := a \cdot \alpha(x) \quad \text{for all } (a, \alpha) \in L \times_Q T, \ x \in L.$$

(3) The map $T \to L \times_Q T; \ \alpha \mapsto (1, \alpha)$ is a monomorphism. The image will be denoted by $1 \times T$. It is the stabilizer of 1 for the above described action of $L \times_Q T$ on L.

(4) The set $L \times 1 := \{(a, 1); a \in L\}$ is a transversal of the coset space $L \times_Q T / 1 \times T$. The map $L \to L \times 1; \ a \mapsto (a, 1)$ is an isomorphism of left loops.

The *Proof* of (1) and (2) can be done simultaneously by looking at the map

$$\Phi : \begin{cases} L \times_Q T \to S_L \\ (a, \alpha) \mapsto \lambda_a \alpha \end{cases}.$$

For all $(a, \alpha), (b, \beta) \in L \times_Q T$, $x \in L$, we compute

$$\Phi(a, \alpha)\Phi(b, \beta)(x) = a \cdot \alpha(b \cdot \beta(x)) = a \cdot \lambda_{\alpha(b)} \lambda_{\alpha(b)}^{-1} \alpha \lambda_b \alpha^{-1} \alpha\beta(x)$$

$$= a \cdot \left(\alpha(b) \cdot \chi(b, \alpha)\alpha\beta(x) \right)$$

$$= \left(a \cdot \alpha(b) \right) \cdot \delta_{a, \alpha(b)} \chi(b, \alpha)\alpha\beta(x)$$

$$= \Phi\left(a \cdot \alpha(b), \, \delta_{a, \alpha(b)} \chi(b, \alpha)\alpha\beta \right)(x).$$

Therefore, Φ is a homomorphism. Φ is injective: if $(a, \alpha)(x) = (b, \beta)(x)$ for all $x \in L$, then by substituting $x = 1$ we find $a = b$, and then $\alpha = \beta$.

The verification of the given inverse is a straightforward, though tedious calculation. Note that it's sufficient to show that it is a right inverse, since we know already that (a, α) is invertible.

We have thus proved that $L \times_Q T$ is in fact a group, which acts on L in the described way.

(3) By definition, the map is a monomorphism, and $1 \times T$ is contained in the stabilizer of 1. Let $(a, \alpha) \in L \times_Q T$ with $1 = a \cdot \alpha(1) = a$, hence $a = 1$ and (a, α) is in the image of our map.

(4) Clearly, $L \times 1$ is a transversal. Denote the corresponding section by σ, and the multiplication on $L \times 1$ by "$\circ$". For $a, b \in L$, $\alpha \in T$ we compute

$$(a, 1)(b, 1)(1, \alpha) = (ab, \delta_{a,b}\alpha) = (ab, 1)(1, \delta_{a,b}\alpha) \in (ab, 1)(1 \times T),$$

therefore

$$(a, 1) \circ (b, 1) = \sigma\big((a, 1)(b, 1)(1 \times T)\big) = (ab, 1).$$

We conclude that the given map is a homomorphism, which is obviously bijective.   ∎

The group $L \times_Q T$ of the preceding theorem will be called the *quasidirect product* of L with T. Frequently, we'll write L and T rather than $L \times 1, 1 \times T$, respectively. The (natural) action of $L \times_Q T$ on L will be used without mention.

As an immediate corollary we get the promised direct construction of $\mathcal{M}_\ell$. Indeed, using notation from the proof of (1), (2.6.2) shows $\Phi(L \times_Q \mathcal{D}) = \mathcal{M}_\ell$. This gives

(2.14) *For every left loop L, the map $L \times_Q \mathcal{D} \to \mathcal{M}_\ell$; $(a, \alpha) \mapsto \lambda_a \alpha$ is an isomorphism.*   ∎

The multiplication in the quasidirect product becomes particularly simple — and more natural — when the group T is sitting inside Aut L. In this case (see (2.12.3)) the map χ becomes redundant, the "obstruction" disappears. Since $\mathcal{D}$ is a subgroup of T, this can only happen, when L is A_ℓ. Thus we have the corollary

(2.15) *Let $(L, \cdot)$ be a left loop, and let T be a transassociant of L, with $T \subseteq$ Aut L. Then L is A_ℓ, and for all $(a, \alpha), (b, \beta) \in L \times_Q T$, we have*

$$(a, \alpha)(b, \beta) = \big(a \cdot \alpha(b), \delta_{a,\alpha(b)}\alpha\beta\big),$$

and

$$(a, \alpha)^{-1} = \big(\alpha^{-1}(a'), \alpha^{-1}\delta_{a,a'}^{-1}\big),$$

where $a' = \lambda_a^{-1}(1)$ is the right inverse of a in L.   ■

Transassociants pop up naturally when we deal with transversals.

(2.16) Theorem. Let G be a group with a subgroup Ω and a transversal L. Let $\theta : G \to S_L$ be the natural permutation representation of G on L. Then $\theta(\Omega)$ is a transassociant of L, and the map $L \times_Q \theta(\Omega) \to \theta(G); (a, \alpha) \mapsto \lambda_a \alpha$ is an isomorphism. Thus the permutation groups $(L \times_Q \theta(\Omega), L)$ and $(\theta(G), L)$ are equivalent.

Proof. First observe that $\theta(\Omega)$ is the stabilizer of 1 in $\theta(G)$. For $a \in L, \omega \in \Omega$, we have

$$\chi(a, \theta_\omega) = \lambda_{\theta_\omega(a)}^{-1} \theta_\omega \lambda_a \theta_\omega^{-1}$$

so 1 is clearly fixed. Since $\lambda_a = \theta_a$ for all $a \in L$, we can conclude $\chi(a, \theta(\Omega)) \subseteq \theta(\Omega)$. Therefore, $\theta(\Omega)$ is a χ-invariant subgroup of $\theta(G)$.

From (2.8.3) we derive that $\mathcal{D} \subseteq \theta(\Omega)$, thus $\theta(G)$ is a transassociant, and $L \times_Q \theta(\Omega)$ is well-defined.

For all $(a, \alpha), (b, \beta) \in L \times_Q \theta(\Omega)$ we have

$$\lambda_a \alpha \lambda_b \beta = \lambda_a \lambda_{\alpha(b)} \lambda_{\alpha(b)}^{-1} \alpha \lambda_b \alpha^{-1} \alpha \beta = \lambda_{a \cdot \alpha(b)} \delta_{a, \alpha(b)} \chi(b, \alpha) \alpha \beta,$$

so the given map is a homomorphism. It is injective, since its kernel is clearly trivial (see also the proof of (2.13)).

For every element $g \in G$ there are (unique) $a \in L$ and $\omega \in \Omega$ such that $g = a\omega$. Hence $\theta_g = \theta_a \theta_\omega = \lambda_a \theta_\omega$, by (2.8.2). Therefore θ_g has the preimage $(a, \theta_\omega) \in L \times_Q \theta(\Omega)$ and the map is surjective.

The last assertion is clear.   ■

If the map θ is injective, i.e., if Ω is corefree, we will also write $L \times_Q \Omega$ instead of the more clumsy $L \times_Q \theta(\Omega)$.

Remarks. 1. (2.14) has been used in many different places to construct $\mathcal{M}_\ell$ for A$_\ell$-loops, e.g., [67; §2], [74; p. 28], [88].

2. The semidirect product for groups is a quasidirect product if the action is faithful. However, the quasidirect product is not a generalization of the semidirect product, since the direct product, for instance, is not a special case of the quasidirect product.

3. KINYON and JONES [73] have introduced a common generalization of the quasidirect product and the semidirect product for groups. Their conditions for the construction to work are rather complicated.

4. With G, L, Ω and θ as in the preceding theorem, let $T := \{\alpha \in S_L;\ \alpha(1) = 1\}$ be the stabilizer of 1 in S_L. It's easy to see that T is a transassociant, and that $S_L \cong L \times_Q T$. Therefore θ induces a homomorphism $\theta' : G \to L \times_Q T$ such that $\theta'(a) = (a, \mathbf{1})$ for all $a \in L$, and $\theta'|_L$ induces an isomorphism of the loops. Notice that $L \times_Q T$ is universal with these properties. This is the content of [83].

5. With G, L, and Ω as in the preceding theorem, the construction of the quasidirect product can be generalized, if L is Ω invariant. Then $L \times \Omega$ can be made into a group isomorphic to G by putting

$$(a, \alpha)(b, \beta) := \big(a \circ \hat{\alpha}(b),\ d_{a,\hat{\alpha}(b)}\alpha\beta\big) \quad \text{for all } (a, \alpha), (b, \beta) \in L \times \Omega,$$

where "$\circ$" denotes the loop multiplication in L. The isomorphism is given by the map $(a, \alpha) \mapsto a\alpha$. The details will be left to the reader.[4] We emphasize that this is only more general, when Ω is not corefree.

---

[4] For the proof of associativity it is helpful to note that $d_{a,\hat{\alpha}(b)}\alpha\beta = (a \circ \hat{\alpha}(b))^{-1}a\alpha b\beta$. This follows from (2.8).

3. The Left Inverse Property and Kikkawa Loops

A. The Left Inverse Property

The following lemma gives an important characterization of the left inverse property. The material is taken from [20; VII.1] and [81; (2.6)].

(3.1) Let L be a groupoid.

(1) The following are equivalent

   (I) L satisfies the left inverse property;

  (II) $\forall a \in L$ there exists a (unique) inverse $a^{-1} \in L$ and $\lambda_{a^{-1}} = \lambda_a^{-1}$;

 (III) $\forall a \in L : \lambda_a^{-1} \in \lambda(L)$;

 (IV) $\forall a \in L$ there exists $a' \in L$ with $\lambda_a \lambda_{a'} = 1$;

  (V) L is a left loop, and $\forall a \in L$ there exists $a' \in L$ such that $a'a = 1$ and $\delta_{a',a} = 1$;

 (VI) L is a left loop, and $\forall a, a' \in L$ with $aa' = 1$ we have $\delta_{a,a'} = 1$;

(VII) L is a left loop with unique inverses, and $\forall a, b \in L$ we have $(ab)^{-1} = \delta_{a,b}(b^{-1}a^{-1})$.

(2) The conditions in (1) imply that the unique solution of the equation $ax = b$, $a, b \in L$, is $x = a^{-1}b$.

Proof. (1) We follow the scheme to the right:

$(I) \rightarrow (II)$

$(III) \leftarrow (VI) \qquad (VII)$

$(V) \leftarrow (IV)$

$(I) \Longrightarrow (II)$: For $a \in L$, let a' be a left inverse of a with $a' \cdot ax = x, \forall x \in L$. This equation is equivalent to $\lambda_{a'} \lambda_a = 1$. Hence λ_a is injective for all $a \in L$.

Now $a' \cdot aa' = a' = a'1$ implies $aa' = 1$. If a'' is another right inverse of a, then $a' = a' \cdot aa'' = a''$, and $a' = a^{-1}$ is the (unique) inverse of a. The equation $\lambda_{a^{-1}} \lambda_a = 1$ also implies that $\lambda_{a^{-1}}$ is surjective, hence bijective.

(II) $\Longrightarrow$ (VII): L is a left loop (with unique inverses), since λ_a is invertible for all $a \in L$. Now, for all $a, b \in L$ we have

$$ab \cdot \delta_{a,b}(b^{-1}a^{-1}) = a(b \cdot b^{-1}a^{-1}) = a(\lambda_b \lambda_{b^{-1}}(a^{-1})) = aa^{-1} = 1,$$

and the result follows.

(VII) $\Longrightarrow$ (IV): Let $a, x \in L$. Since L is a left loop, we have

$$1 = x^{-1}a \cdot \delta_{x^{-1},a}(a^{-1}x) = x^{-1}(a \cdot a^{-1}x) = x^{-1}\lambda_a\lambda_{a^{-1}}(x).$$

This implies $\lambda_a\lambda_{a^{-1}}(x) = x$, therefore $\lambda_a\lambda_{a^{-1}} = \mathbf{1}$.

(IV) $\Longrightarrow$ (V): From $\lambda_a\lambda_{a'} = \mathbf{1}$ we derive that λ_a is surjective, and $\lambda_{a'}$ is injective for all $a \in L$. Therefore $\lambda_{a'}$ is bijective and $\lambda_a = \lambda_{a'}^{-1}$. Thus L is a left loop. The other statements are easy.

(V) $\Longrightarrow$ (III): For every $x \in L$ we have

$$\lambda_{a'}\lambda_a(x) = a' \cdot ax = a'a \cdot \delta_{a',a}(x) = x, \quad \text{thus} \quad \lambda_{a'}\lambda_a = \mathbf{1}.$$

Since λ_a is bijective, we can conclude that $\lambda_a^{-1} = \lambda_{a'} \in \lambda(L)$.

(III) $\Longrightarrow$ (I): For $a' \in L$ with $\lambda_a^{-1} = \lambda_{a'}$ we have $a' \cdot ax = \lambda_{a'}\lambda_a(x) = x$.

(II) $\Longrightarrow$ (VI): Clearly, $a' = a^{-1}$. Therefore $\delta_{a,a'} = \delta_{a,a^{-1}} = \lambda_{aa^{-1}}^{-1}\lambda_a\lambda_{a^{-1}} = \lambda_a\lambda_a^{-1} = \mathbf{1}$.

(VI) $\Longrightarrow$ (III): $\mathbf{1} = \delta_{a,a'} = \lambda_{aa'}^{-1}\lambda_a\lambda_{a'} = \lambda_a\lambda_{a'}$. By hypothesis λ_a is bijective, hence $\lambda_a^{-1} = \lambda_{a'} \in \lambda(L)$.

(3) To solve the equation $ax = b$, we simply multiply it by a^{-1} on both sides. $\blacksquare$

The proofs also show all but the last statement of

(3.2) *Let L be a left loop and $a \in L$. Then a satisfies the left inverse property if and only if there exists $a' \in L$ with $\lambda_{a'} = \lambda_a^{-1}$. If this is the case, then a has a unique inverse $a' = a^{-1}$, and for all $b \in L$ we have*

$$\delta_{a^{-1},ab}\delta_{a,b} = \mathbf{1}.$$

Proof of the last statement.

$$\delta_{a^{-1},ab}\delta_{a,b} = \lambda_{a^{-1}\cdot ab}^{-1}\lambda_{a^{-1}}\lambda_{ab}\lambda_{ab}^{-1}\lambda_a\lambda_b = \lambda_b^{-1}\lambda_a^{-1}\lambda_a\lambda_b = \mathbf{1}. \quad \blacksquare$$

Next we give a characterization of the left inverse property in terms of transversals. The second part was inspired by [81; (3.9),(3.12)].

(3.3) *Let* $(L, \circ)$ *be a transversal of the coset space* G/Ω. *If* $L^{-1} \subseteq L$, *where the inverses are taken in* G, *then we have*

(1) L *satisfies the left inverse property, and the inverses of elements of* L *formed in* $(L, \circ)$ *and in* G *coincide.*

(2) *The following are equivalent*

    (I) L *satisfies the automorphic inverse property;*

    (II) *For all* $a, b \in L : abd_{a,b}^{-1} = d_{a^{-1},b^{-1}}ba$;

    (III) *For all* $a, b \in L$ *there exists* $\omega \in \Omega$ *such that* $abd_{a,b}^{-1} = \omega ba$.

Proof. (1) We shall use (2.8). In particular, let θ be the natural permutation representation of G on L.

For $a \in L$ we have $a^{-1} \in L$, therefore, $\lambda_{a^{-1}} = \theta_{a^{-1}} = \theta_a^{-1} = \lambda_a^{-1}$, and (3.1.1) shows the result.

(2) (I) $\Longrightarrow$ (II): $abd_{a,b}^{-1} = a \circ b = (a^{-1} \circ b^{-1})^{-1} = (a^{-1}b^{-1}d_{a^{-1},b^{-1}}^{-1})^{-1} = d_{a^{-1},b^{-1}}ba$.

"(II) $\Longrightarrow$ (III)" is trivial.

(III) $\Longrightarrow$ (I): $(a \circ b)^{-1} = (abd_{a,b}^{-1})^{-1} = (\omega ba)^{-1} = a^{-1}b^{-1}\omega^{-1} \in L$. Since also $a^{-1} \circ b^{-1} = a^{-1}b^{-1}d_{a^{-1},b^{-1}}^{-1} \in L$, we must have $\omega = d_{a^{-1},b^{-1}}$ and $(a \circ b)^{-1} = a^{-1} \circ b^{-1}$. ∎

Fixed point free transassociants will be important in §7. We record some simple consequences when $\mathcal{D}$ is fixed point free.

(3.4) *Let* L *be a left loop such that* $\mathcal{D}(L)$ *acts fixed point free on* $L^{\#}$. *If every element of* L *has unique inverses, i.e.,* $ab = 1 \Longrightarrow ba = 1$, *then* L *satisfies the left inverse property. If, in addition,* L *is automorphic inverse, then* L *is also left alternative.*

Proof. Let $a \in L^{\#}$. We have $a = a \cdot a^{-1}a = aa^{-1} \cdot \delta_{a,a^{-1}}(a) = \delta_{a,a^{-1}}(a)$. The hypothesis enforces $\delta_{a,a^{-1}} = 1$. Thus (3.1.1) shows that L satisfies the left inverse property.

Assume L is also automorphic inverse. Then by (3.1.1) we have for all $a \in L$

$$\delta_{a,a}(a^{-1}a^{-1}) = (aa)^{-1} = (a^{-1})^2.$$

If $(a^{-1})^2 \neq 1$ we must have $\delta_{a,a} = 1$, and a is left alternative. If $(a^{-1})^2 = 1$, then $a = a^{-1}$, and $a \cdot ax = x = a^2 x$ by the left inverse property. Therefore every element of L is left alternative.   ∎

The following two lemmas will be useful for the upcoming discussion of Kikkawa loops.

(3.5) *Let L be a left loop, and for $a \in L$ let a^ϱ denote the right inverse, i.e., $aa^\varrho = 1$. Then any two of the following conditions imply the third.*

   (I)   *L satisfies the left inverse property;*

  (II)   *L satisfies the automorphic inverse property;*

 (III)   *$a(ab)^\varrho = b^\varrho$ for all $a, b \in L$.*

Proof. Both, (I) and (III) imply unique inverses, i.e., $a^\varrho a = 1$: For (I) this is in (3.1). For (III) put $b = a^\varrho$ and compute $(a^\varrho)^\varrho = a(aa^\varrho)^\varrho = a$ thus $a^\varrho a = 1$. Therefore we can write $a^{-1} = a^\varrho$.

If (I) and (II) are valid, we can compute $a(ab)^{-1} = a \cdot a^{-1} b^{-1} = b^{-1}$ for all $a, b \in L$.

(I) and (III): $a(ab)^{-1} = b^{-1}$ implies $(ab)^{-1} = a^{-1} b^{-1}$ for $a, b \in L$.

Finally, from (II) and (III) we obtain $a^{-1}(ab) = a^{-1}(a^{-1} b^{-1})^{-1} = b$ for all $a, b \in L$.   ∎

(3.6) *Let L be a left loop with $\lambda^2_{ab} = \lambda_a \lambda^2_b \lambda_a$ for all $a, b \in L$, then for all $a, b \in L$ we have*

(1) *L has unique inverses and $a(ab)^{-1} = b^{-1}$.*

(2) *$ab = \delta_{a,b}(ba)$.*

(3) *The left and automorphic inverse property are equivalent.*

Proof. (1) We have $(ab)^2 = a(b \cdot ba)$ by applying the equation from the hypothesis to 1. Now define a^ϱ by $aa^\varrho = 1$, and compute

$$aa^\varrho = 1 = a(a^\varrho \cdot a^\varrho a) \implies a^\varrho = a^\varrho \cdot a^\varrho a \implies a^\varrho a = 1.$$

Thus $a^\varrho = a^{-1}$. Now we can calculate

$$ab \cdot \left(ab \cdot \lambda_a^{-1}(b^{-1})\right) = \lambda^2_{ab}\left(\lambda_a^{-1}(b^{-1})\right) = \lambda_a \lambda^2_b \lambda_a \left(\lambda_a^{-1}(b^{-1})\right) = ab.$$

Canceling ab and rearranging yields $(ab)^{-1} = \lambda_a^{-1}(b^{-1})$, which is equivalent with the second assertion.

(2) From the assumption we get $\lambda_{ab} = \delta_{a,b}\lambda_b\lambda_a$. Apply this to 1 for the result.

(3) follows directly from (1) and (3.5). ∎

Remarks. 1. The identity (III) of (3.5) (in fact its dual) has been studied by JOHNSON and SHARMA [51].

2. (3.6.3) is due to KINYON and JONES [73; Thm. 4.1.1].

B. KIKKAWA LEFT LOOPS

We'll now deal with the automorphic inverse property. An A_ℓ-left-loop with the left and automorphic inverse property will be called a *Kikkawa left loop*, and a *Kikkawa loop* if it is a loop.

(3.7) Theorem. *Let L be a left loop with left inverse property.*

(1) *If L satisfies the automorphic inverse property, then for all $a, b \in L$ the following are equivalent*

    (I)   The map $\iota : x \mapsto x^{-1}$ commutes with $\delta_{a,b}$, i.e.,
$$\iota\delta_{a,b} = \delta_{a,b}\iota;$$

    (II) $\delta_{a,b} = \delta_{a^{-1},b^{-1}};$

    (III) $\lambda_{ab}^2 = \lambda_a\lambda_b^2\lambda_a.$

(2) *If L is an A_ℓ-left-loop, then for all $a, b \in L$ we have*

$$\delta_{a,b} = \delta_{b^{-1},a^{-1}}^{-1} = \delta_{b,b^{-1}a^{-1}} .$$

(3) *If L is a Kikkawa left loop, then the identities in (1) are satisfied for all $a, b \in L$, and*

$$\delta_{a,b}^{-1} = \delta_{b,a} \quad \text{for all } a, b \in L.$$

Proof. (1) (I) $\Longleftrightarrow$ (II): ι is an automorphism of L, therefore, using (2.4.2)
$$\iota\delta_{a,b}\iota = \delta_{\iota(a),\iota(b)} = \delta_{a^{-1},b^{-1}}.$$

This is equal to $\delta_{a,b}$ if and only if ι commutes with $\delta_{a,b}$, since ι is an involution.

(II) $\Longleftrightarrow$ (III): Using the left and automorphic inverse property, we obtain

$$\delta_{a,b} = \delta_{a^{-1},b^{-1}} \iff \lambda_{ab}^{-1}\lambda_a\lambda_b = \lambda_{a^{-1}b^{-1}}^{-1}\lambda_{a^{-1}}\lambda_{b^{-1}} = \lambda_{ab}\lambda_a^{-1}\lambda_b^{-1}$$
$$\iff \lambda_{ab}^2 = \lambda_a\lambda_b^2\lambda_a.$$

(2) Using (2.4.1) and (3.1.1) we can compute

$$\delta_{a,b}\lambda_{b^{-1}a^{-1}}\delta_{a,b}^{-1} = \lambda_{\delta_{a,b}(b^{-1}a^{-1})} = \lambda_{(ab)^{-1}} = \lambda_{ab}^{-1}.$$

Rewriting this, and applying (3.1.1) repeatedly, we obtain

$$\begin{aligned}
\delta_{a,b} &= \lambda_{ab}\delta_{a,b}\lambda_{b^{-1}a^{-1}} = \lambda_a\lambda_b\lambda_{b^{-1}a^{-1}} \\
&= \lambda_{a^{-1}}^{-1}\lambda_b\lambda_{b^{-1}a^{-1}} = \lambda_{b\cdot b^{-1}a^{-1}}^{-1}\lambda_b\lambda_{b^{-1}a^{-1}} = \delta_{b,b^{-1}a^{-1}} \\
&= \lambda_{a^{-1}}^{-1}\lambda_{b^{-1}}^{-1}\lambda_{b^{-1}a^{-1}} = \delta_{b^{-1},a^{-1}}^{-1}.
\end{aligned}$$

These are the claimed identities.

(3) If L is A_ℓ, then clearly ι centralizes $\mathcal{D}(L)$. Thus for a Kikkawa left loop (1)(I) is satisfied. Combining (II) and (2) yields the last identity. ∎

Remarks. 1. In the proof of "(I) $\Longleftrightarrow$ (II)", the left inverse property is not really needed, it suffices to have unique inverses.

2. The material has been generalized from [67; Prop. 1.13 and Lemma 1.8]. "Kikkawa loop" is our terminology, KIKKAWA called them *symmetric loops*.

3. There do exist examples of Kikkawa left loops, which are not loops, see (12.3.2).

The following lemma is due to KREUZER [77; (1.4)]. We use basically his proof, but with weaker hypotheses.

(3.8) Let L be a loop which satisfies the identity $(ab)^2 = a\cdot b^2 a$ for all $a, b \in L$. Then the map $\kappa : L \to L;\ x \mapsto x^2$ is injective if and only if L contains no elements of order 2, i.e., $a^2 = 1 \implies a = 1$ for all $a \in L$.

In particular, if L is a Kikkawa loop, then the general hypothesis is true if and only if L left alternative.

Proof. If κ is injective, then there is clearly only one element $a \in L$ with $a^2 = 1$, namely $a = 1$.

Assume for the converse that $\kappa(a) = \kappa(b)$, $a, b \in L$. There exists $c \in L$ with $ac = b$, thus

$$a^2 = b^2 = (ac)^2 = a \cdot c^2 a \implies c^2 = 1 \implies c = 1.$$

Therefore $a = b$, and κ is injective.

Now assume that L is a Kikkawa loop, then for all $a, b \in L$ we have $\lambda_{ab}^2 = \lambda_a \lambda_b^2 \lambda_a$ by (3.7.3). Applying this to 1 gives $(ab)^2 = a(b \cdot ba)$, which is equal to $a \cdot b^2 a$ if and only if $b \cdot ba = b^2 a$. Hence the last assertion. ∎

Remark. A group satisfies the identity $(ab)^2 = a \cdot b^2 a$ if and only if it is commutative. Of course, for commutative groups the conclusion of the lemma is well-known. It is even true for periodic groups, i.e., groups such that every element is of finite order.

C. The Bol Condition

Recall that a groupoid L is called Bol if

$$a(b \cdot ac) = (a \cdot ba)c \quad \text{for all} \quad a, b, c \in L.$$

We begin with an almost trivial observation.

(3.9) *For a groupoid L the following are equivalent*

    (I) L *is Bol;*

    (II) $\lambda_a \lambda_b \lambda_a = \lambda_{a \cdot ba}$ *for all $a, b \in L$;*

    (III) $\lambda_a \lambda(L) \lambda_a \subseteq \lambda(L)$ *for all $a \in L$.*

Proof. "(I) $\implies$ (II)" and "(II) $\implies$ (III)" are trivial.

(III) $\implies$ (I): For $a, b \in L$ we have $\lambda_a \lambda_b \lambda_a = \lambda_c$ for suitable $c \in L$. Applying this to 1 yields $c = a \cdot ba$. Thus we can compute for all $x \in L$

$$a(b \cdot ax) = \lambda_a \lambda_b \lambda_a(x) = \lambda_{a \cdot ba}(x) = (a \cdot ba)x,$$

and L is Bol. ∎

The following lemma will be useful, when we construct examples later.

(3.10) *Let L be a Bol groupoid.*

(1) *L is left alternative.*

(2) *If every $a \in L$ has a right inverse a^ϱ, then a^ϱ is also a left inverse of a.*

(3) *If every $a \in L$ has a left inverse a', then L is a Bol loop. In particular, L satisfies the left inverse property, and for all $a, b \in L$, we have*

$$ax = b \iff x = a^{-1}b \quad \text{and}$$

$$ya = b \iff y = a^{-1}(ab \cdot a^{-1}), \quad \text{where } a^{-1} = a'.$$

(4) *If L is a Bol loop and U a non-empty subset of L subject to the condition $\forall a, b \in U : ab \in U, a^{-1} \in U$, then U is a Bol subloop of L.*

Proof. (1) Put $b = 1$ in $a(b \cdot ac) = (a \cdot ba)c$.

(2) $a^\varrho a = a^\varrho(a \cdot a^\varrho a^{\varrho\varrho}) = (a^\varrho \cdot aa^\varrho)a^{\varrho\varrho} = a^\varrho a^{\varrho\varrho} = 1$.

(3) Our first aim is to prove the left inverse property:

$$\lambda_{a'}\lambda_{a''}\lambda_{a'} = \lambda_{a' \cdot a''a'} = \lambda_{a'} \quad \text{where} \quad a'' = (a')'.$$

Apply this to a to obtain $a'a'' = 1$. Thus using (1), Bol, and (1) again, we can compute

$$\lambda_{a''}\lambda_{a'}\lambda_{a'}\lambda_{a''} = \lambda_{a''}\lambda_{(a')^2}\lambda_{a''} = \lambda_{a'' \cdot (a')^2 a''} = \lambda_{a''(a' \cdot a'a'')} = 1.$$

Applying $\lambda_{a''}\lambda_{a'}$ on the left and using the two displayed equations above yields

$$\lambda_{a''}\lambda_{a'} = \lambda_{a''}\,\lambda_{a'}\lambda_{a''}\lambda_{a'}\,\lambda_{a'}\lambda_{a''} = \lambda_{a''}\lambda_{a'}\lambda_{a'}\lambda_{a''} = 1.$$

Now $a'' = \lambda_{a''}\lambda_{a'}(a) = a$, and so $\lambda_a\lambda_{a'} = 1$. Therefore (3.1) shows that L is a left loop, that we can write $a' = a^{-1}$, and that the first equivalence of the last statement is true.

To prove the second, we compute

$$a^{-1}(ab \cdot a^{-1}) \cdot a = a^{-1}(ab \cdot a^{-1}a) = a^{-1} \cdot ab = b,$$

thus y is a solution. If y_1 is another solution, then $ya = y_1 a$ and using the Bol identity we infer

$$ay = a(y \cdot aa^{-1}) = (a \cdot ya)a^{-1}$$
$$= (a \cdot y_1 a)a^{-1} = a(y_1 \cdot aa^{-1}) = ay_1.$$

Since L is a left loop, $y = y_1$, and the solution is unique. So L is actually a loop.

(4) Clearly U is a groupoid, and in particular contains 1. The hypotheses from (2) are inherited by U, so (2) gives the result. ∎

As an immediate corollary we get

(3.11) *A left loop with the Bol condition is a Bol loop.* ∎

Remarks. 1. A very similar, but slightly weaker result can be found in [109; Thm. 2]. Here a left loop has been used, which is only required to have a right identity ε. Variations of this are [81; (2.13)], and under even stronger hypothesis [40; Lemma 2]. In [25] (see also [102]) it is shown that a Bol quasigroup always has a right identity.

2. Notice also that $(\mathbf{Z}, -)$, the integers with the binary operation $(a, b) \mapsto a - b$, is a Bol quasigroup.

3. In this context it seems worthwhile to note that a quasigroup satisfying any of the Moufang identities necessarily has an identity, hence is a Moufang loop (cf. [82]).

4. The solution y from (3) in (3.10) occurs in [10; VI.6.8 p. 106] and [40; Lemma 2].

5. The proof of (3.10.3) becomes much easier if the hypothesis "all the λ_a are injective" is added. The proof of the more general statement is due to KINYON [70].

6. Using methods from §6, one can show that in every Bol groupoid L we have $\lambda_{a^k} = \lambda_a^k$ for all $a \in L$, $k \in \mathbf{N}$.

As an important application, we establish a condition for a transversal to be a Bol loop.

(3.12) *Let L be a transversal of the coset space G/Ω, and let N be the core of Ω in G. If $aLa \subseteq LN$ for all $a \in L$, then L is a Bol loop.*

Proof. Let θ be the natural permutation representation of G on L (see (2.8)). The kernel of θ is N. For $a, b \in L$ there exists $n \in N$ with $aban \in L$. Therefore

$$\lambda_a \lambda_b \lambda_a = \theta_a \theta_b \theta_a = \theta_{aba} = \theta_{aban} = \lambda_{aban} \in \lambda(L).$$

Thus (3.9) and (3.11) show the result, since L is a left loop. ∎

4. Isotopy Theory

We'll develop isotopy theory, firstly, to give a systematic exposition of some parts, which seem to be missing in the literature, secondly, because of its applications to Bol loops. The definition is most general, but most theorems will be formulated only for loops, or left loops.

Let L, L' be groupoids. A triple $(\alpha, \beta, \gamma) : L \to L'$ of bijective maps is called an *isotopism* if

$$\alpha(x)\beta(y) = \gamma(xy) \quad \text{for all } x, y \in L.$$

A *principal isotopism* is an isotopism with $\gamma = 1$, in particular, L and L' are defined on the same underlying set. If a (principal) isotopism $L \to L'$ exists, then L' is called a (*principal*) *isotope* of L. In case of $L = L'$, we speak of *autotopisms*. Clearly, the set $\mathrm{Top}(L)$ of all autotopisms of L forms a group under componentwise composition. In fact, it is a subgroup of $\mathcal{S}_L^3$, the threefold cartesian product of the symmetric group on L with itself.

Obviously, every isomorphism $\alpha : L \to L'$ gives an isotopism (α, α, α). At occasions, we shall identify these two. Conversely, an isotopism with three equal components defines naturally an isomorphism. In this sense, $\mathrm{Aut}(L)$ is a subgroup of $\mathrm{Top}(L)$.

(4.1) *Let L be a left loop and $a, b \in L$ such that ϱ_b is invertible. Put $x \circ y := \varrho_b^{-1}(x)\lambda_a^{-1}(y)$ for all $x, y \in L$. Then $L^{(a,b)} := (L, \circ)$ is a left loop with identity ab, and $(\varrho_b^{-1}, \lambda_a^{-1}, 1) : L^{(a,b)} \to L$ is a principal isotopism. If L is a loop, then so is $L^{(a,b)}$. Moreover, $\mathcal{M}_\ell(L^{(a,b)}) = \mathcal{M}_\ell(L)$.*

A *Proof* is needed only for the last statement: The left translation $y \mapsto u \circ y, u, y \in L$, of $L^{(a,b)}$ is given by $\lambda_{\varrho_b^{-1}(u)}\lambda_a^{-1}$, which is in $\mathcal{M}_\ell(L)$. Therefore $\mathcal{M}_\ell(L^{(a,b)}) \subseteq \mathcal{M}_\ell(L)$, by symmetry, $\mathcal{M}_\ell(L^{(a,b)}) = \mathcal{M}_\ell(L)$. Note that the inverse of $(\varrho_b^{-1}, \lambda_a^{-1}, 1)$ is a principal isotopism $L \to L^{(a,b)}$. ∎

The principal isotopes described in the theorem are, up to isomorphism, all principal isotopes of L. In fact, putting $y = 1$ and then $x = 1$ in the defining equations of an isotopism, one easily derives

(4.2) *Let* L, L' *be left loops and let* $(\alpha, \beta, \gamma) : L' \to L$ *be an isotopism. For* $a := \alpha(1)$ *and* $b := \beta(1)$ *we have*

(1) ϱ_b *is invertible,* $\alpha = \varrho_b^{-1}\gamma$ *and* $\beta = \lambda_a^{-1}\gamma$. *Here* λ *and* ϱ *correspond to the multiplication in* L.

(2) (α, β, γ) *factors into the principal isotopism* $(\varrho_b^{-1}, \lambda_a^{-1}, 1)$: $L^{(a,b)} \to L$ *and the isomorphism* $\gamma : L' \to L^{(a,b)}$. *Indeed* $(\alpha, \beta, \gamma) = (\varrho_b^{-1}, \lambda_a^{-1}, 1)(\gamma, \gamma, \gamma)$. ∎

The second part of this theorem is often expressed by saying that every isotope of a left loop L is isomorphic to a principal isotope of L. Therefore many questions about isotopes can be reduced to questions about principal isotopes. We'll make use of this soon.

The statement of the last paragraph is true in a much more general context (cf. [20; III.1 p. 56f]).

The first part of the theorem implies that two components of an isotopism determine the third. This will occasionally be used by leaving out a component of an isotopism, e.g., $(\,.\,, \beta, \gamma)$ refers to the unique isotopism $(\varrho_{\beta(1)}^{-1}\gamma, \beta, \gamma)$ if we know the existence beforehand.

For a more precise statement in the case of autotopisms, we introduce the projection to the i-th coordinate

$$\pi_i : \mathrm{Top}(L) \to \mathcal{S}_L; \ (\alpha_1, \alpha_2, \alpha_3) \mapsto \alpha_i, \quad i \in \{1, 2, 3\}.$$

These mappings will be used later on. We have

(4.3) *Let* L *be a left loop. The maps*

$$\pi_i \times \pi_j : \begin{cases} \mathrm{Top}(L) \to \mathcal{S}_L^2 \\ (\alpha_1, \alpha_2, \alpha_3) \mapsto (\alpha_i, \alpha_j) \end{cases} \quad \text{for } 1 \leq i < j \leq 3$$

are monomorphisms.

Proof. Obviously, $\pi_i \times \pi_j$ is a homomorphism. By (4.2.1), the kernel is trivial. ∎

In case of autotopisms, the symbol $(\,.\,, \beta, \gamma)$ can be viewed as a shorthand for $(\pi_2 \times \pi_3)^{-1}(\beta, \gamma)$.

A permutation $\beta : L \to L$ is called a *(left) pseudoautomorphism* if there exists $b \in L$ such that

$$b\beta(x) \cdot \beta(y) = b\beta(xy) \quad \text{for all } x, y \in L.$$

b is called a *companion* of β. The set of all pseudoautomorphisms will be denoted by $\Psi(L)$. The dual notion is called *right pseudoautomorphism*. We shall first concentrate on pseudoautomorphisms.

(4.4) *Let L be a left loop.*

(1) *For $b \in L$ and a permutation β of L the following are equivalent*

   (I) $\beta \in \Psi(L)$ *with companion b;*

   (II) $(\lambda_b\beta, \beta, \lambda_b\beta) \in \text{Top } L$;

   (III) $\beta\lambda_x\beta^{-1} = \lambda_b^{-1}\lambda_{b\beta(x)}$ *for all $x \in L$.*

(2) *If $\beta \in \Psi(L)$, then $\beta(1) = 1$.*

(3) *If $(\alpha, \beta, \gamma) \in \text{Top } L$, then*

$$\beta(1) = 1 \iff \alpha = \gamma \iff \beta \in \Psi(L).$$

In this case, a companion of β is $\alpha(1) = \gamma(1)$.

(4) $\text{T}\Psi(L) := \{(\alpha, \beta, \gamma) \in \text{Top}(L); \ \beta(1) = 1\}$ *is a subgroup of* $\text{Top}(L)$. *The map*

$$\pi_2|_{\text{T}\Psi(L)} : \begin{cases} \text{T}\Psi(L) \to \Psi(L) \\ (\alpha, \beta, \gamma) \mapsto \beta \end{cases}$$

is an epimorphism.

(5) $\Psi(L)$ *is a group.*

Proof. (1) "(I) $\iff$ (II)" is just a rewording of the definitions.

(I) $\iff$ (III): For all $x, y \in L$ we have

$$b\beta(xy) = b\beta(x) \cdot \beta(y) \iff \lambda_b\beta\lambda_x(y) = \lambda_{b\beta(x)}\beta(y)$$

$$\iff \beta\lambda_x\beta^{-1} = \lambda_b^{-1}\lambda_{b\beta(x)}.$$

(2) In the equation $b\beta(1) \cdot \beta(1) = b\beta(1)$, we can first cancel $\beta(1)$ on the right by (1) and (4.2.1). Then we cancel b on the left, to obtain $\beta(1) = 1$.

(3) If $\beta(1) = 1$, then $\alpha = \gamma$ by (4.2.1).

If $\alpha = \gamma$, then $(\alpha, \beta, \gamma) = (\lambda_{\alpha(1)}\beta, \beta, \lambda_{\alpha(1)}\beta)$ by (4.2.1), again. So $\beta \in \Psi(L)$ with companion $\alpha(1)$ by (1).

If $\beta \in \Psi(L)$ then $\beta(1) = 1$ by (2).

(4) Clearly, $\mathrm{T}\Psi(L)$ is a group. By (3) $\pi_2\big(\mathrm{T}\Psi(L)\big) \subseteq \Psi(L)$, so π_2 is well-defined. Since π_2 is the restriction of a projection, it is a homomorphism, which is surjective by (1) and (2).

(5) follows immediately from (4). ∎

Remarks. **1.** Using (1)(III), it's straightforward to compose a direct proof of (5).

2. Pseudoautomorphisms can be characterized using the obstruction map χ from §2.C. Indeed, from (1)(III) one derives $\beta \in \Psi(L)$ with companion b if and only if $\chi(x, \beta) = \delta_{b,\beta(x)}^{-1}$ for all $x \in L$. See [73; Prop. 2.1].

3. (3) is a generalization of [10; Lemma 5.5 p. 84].

4. In the literature frequently right pseudoautomorphisms are used, e.g., [20], [10], [94]. Our choice fits better into our future needs. Still right pseudoautomorphisms have their place as we shall see in a minute.

Indeed, our aim is to present BELOUSOV's characterization of loops which are isomorphic to every loop isotope. Such loops have been called *G-loops*. To formulate this characterization we need some facts about right pseudoautomorphisms. We shall not completely dualize (4.4), but we record

(4.5) *Let L be a loop.*

(1) *A permutation α of L is a right pseudoautomorphism with companion $a \in L$ if and only if $(\alpha, \varrho_a\alpha, \varrho_a\alpha) \in \mathrm{Top}(L)$.*

(2) *If $(\alpha, \beta, \gamma) \in \mathrm{Top}(L)$, then α is a right pseudoautomorphism if and only if $\alpha(1) = 1$. In this case $\beta(1) = \gamma(1)$ is a companion of α.* ∎

We give a description of isomorphisms between principal isotopes.

(4.6) *Let L be a loop, $a, b, c, d \in L$, and let γ be a permutation of L. Then the following are equivalent*

(I) $\gamma : L^{(a,b)} \to L^{(c,d)}$ is an isomorphism;

(II) $(\varrho_d^{-1}\gamma\varrho_b, \lambda_c^{-1}\gamma\lambda_a, \gamma) \in \text{Top}(L)$ and $\gamma(ab) = cd$;

(III) There exists $(\alpha, \beta, \gamma) \in \text{Top}(L)$ with $\alpha(a) = c$, $\beta(b) = d$, $\gamma(ab) = cd$.

Proof. (I) $\Longrightarrow$ (II): By (4.1)

$$(\varrho_d^{-1}\gamma\varrho_b, \lambda_c^{-1}\gamma\lambda_a, \gamma) = (\varrho_d^{-1}, \lambda_c^{-1}, \mathbf{1})(\gamma, \gamma, \gamma)(\varrho_b, \lambda_a, \mathbf{1}) \in \text{Top}(L).$$

Moreover, since γ is an isomorphism, it maps the identity ab of $L^{(a,b)}$ to the identity cd of $L^{(c,d)}$.

(II) $\Longrightarrow$ (III): Put $(\alpha, \beta, \gamma) = (\varrho_d^{-1}\gamma\varrho_b, \lambda_c^{-1}\gamma\lambda_a, \gamma)$.

(III) $\Longrightarrow$ (I): Using (4.1) again, we find that

$$\phi := (\varrho_d, \lambda_c, \mathbf{1})(\alpha, \beta, \gamma)(\varrho_b^{-1}, \lambda_a^{-1}, \mathbf{1}) : L^{(a,b)} \to L^{(c,d)}$$

is an isotopism. Now every component of this isotopism maps ab, the identity of $L^{(a,b)}$, to cd, the identity of $L^{(c,d)}$. By (4.2.1), all components of ϕ are equal to γ, hence γ is an isomorphism. ∎

We are now ready to give BELOUSOV's theorem [10; Thm. 3.8, p. 48]. His proof is also presented in [94; III.6.1, p. 82]. We'll follow DRISKO's exposition [29; Cor. 4,5], which adds (IV) below to the series of equivalent conditions. The major part of the preceding lemma (4.6) is also his.

(4.7) Theorem. *Let L be a loop, and $a \in L$. For the principal isotopes we have*

(1) L *is isomorphic to* $L^{(a,1)}$ *if and only if a is the companion of a pseudoautomorphism of L.*

(2) L *is isomorphic to* $L^{(1,a)}$ *if and only if a is the companion of a right pseudoautomorphism of L.*

(3) *The following are equivalent*

(I) *Every isotope of L is isomorphic to L (i.e., L is a G-loop);*

(II) L *is isomorphic to the principal isotopes $L^{(a,1)}$ and $L^{(1,a)}$ for every $a \in L$;*

(III) *Every element of L is the companion of a pseudoautomorphism and of a right pseudoautomorphism;*

(IV) $(\pi_1 \times \pi_2)\big(\mathrm{Top}(L)\big)$ *acts transitively on* $L \times L$.

For the *Proof* note that $L = L^{(1,1)}$. By (4.2.2) it suffices to consider principal isotopes.

(1) By (4.6) the map $\gamma : L \to L^{(a,1)}$ is an isomorphism if and only if $(\gamma, \lambda_a^{-1}\gamma, \gamma) \in \mathrm{Top}(L)$, and $\gamma(1) = a$. By (4.4.3) this is equivalent to $\lambda_a^{-1}\gamma$ being a pseudoautomorphism of L with companion a.

(2) By (4.6) the map $\gamma : L \to L^{(1,a)}$ is an isomorphism if and only if $(\varrho_a^{-1}\gamma, \gamma, \gamma) \in \mathrm{Top}(L)$, and $\gamma(1) = a$. Using (4.5) the proof can be completed as before.

(3) "(I) $\Longrightarrow$ (III)" and "(III) $\Longrightarrow$ (II)" are direct consequences of (1) and (2).

(II) $\Longrightarrow$ (IV): Let $a, b \in L$. Because $\mathrm{Top}(L)$ is a group, it suffices to show that there exists

$$(\alpha, \beta, \gamma) \in \mathrm{Top}(L) \quad \text{with} \quad \alpha(1) = a, \ \beta(1) = b.$$

By assumption and (4.6) there exists

$$(\alpha_1, \alpha_2, \alpha_1) \in \mathrm{Top}(L) \quad \text{with} \quad \alpha_1(1) = a, \ \alpha_2(1) = 1.$$

Put $d := \alpha_2^{-1}(b)$. Again by (4.6) there exists

$$(\beta_1, \beta_2, \beta_2) \in \mathrm{Top}(L) \quad \text{such that} \quad \beta_1(1) = 1, \ \beta_2(1) = d.$$

Now we have

$$(\alpha_1, \alpha_2)(\beta_1, \beta_2)(1, 1) = (\alpha_1, \alpha_2)(1, d) = (a, b).$$

(IV) $\Longrightarrow$ (I): For $a, b \in L$, choosing $(\alpha, \beta, \gamma) \in \mathrm{Top}(L)$ with $\alpha(1) = a$, $\beta(1) = b$, then (4.6) shows that $\gamma : L \to L^{(a,b)}$ is an isomorphism, since $\gamma(1) = \alpha(1)\beta(1) = ab$. ∎

Remarks. 1. Wilson proves the equivalence of (I) and (II) in [123].

2. Robinson [101] gives an example of a Bol loop with all loop isotopes isomorphic. The example is described briefly in §12, Example 4.

5. Nuclei and the Autotopism Group

In every groupoid L one can define the *left, middle, right nucleus* of L, respectively,

$$\mathcal{N}_\ell(L) := \{a \in L; \ \forall x, y \in L : a \cdot xy = ax \cdot y\}$$
$$\mathcal{N}_m(L) := \{a \in L; \ \forall x, y \in L : x \cdot ay = xa \cdot y\}$$
$$\mathcal{N}_r(L) := \{a \in L; \ \forall x, y \in L : x \cdot ya = xy \cdot a\}.$$

These are obviously semigroups with 1 (i.e., they are associative groupoids). For convenience, we shall drop the arguments if they are clear from the context, writing briefly $\mathcal{N}_\ell, \mathcal{N}_m, \mathcal{N}_r$.

If L is a loop, then the nuclei can be embedded into the autotopism group. These embedded groups are the kernels of the projections π_i from §4. Recall that $\mathrm{Top}(L)$ is a subgroup of $\mathcal{S}_L \times \mathcal{S}_L \times \mathcal{S}_L$, and π_i is the restriction to $\mathrm{Top}(L)$ of the projection onto the i-th coordinate.

(5.1) *Let L be a left loop, and let $\mathrm{T}\mathcal{N}_\ell$ be the kernel of $\pi_2 : \mathrm{Top}(L) \to \mathcal{S}_L$. We have*

(1) *For $a \in L$, the following are equivalent*

  (I) $a \in \mathcal{N}_\ell$;

  (II) $(\lambda_a, 1, \lambda_a) \in \mathrm{Top}(L)$;

  (III) $\lambda_a \lambda_x = \lambda_{ax}$ *for all $x \in L$;*

  (IV) $\delta_{a,x} = 1$ *for all $x \in L$.*

(2) $\mathrm{T}\mathcal{N}_\ell$ *is a normal subgroup of $\mathrm{Top}(L)$, and the map*

$$\tau_\ell : \mathcal{N}_\ell \to \mathrm{T}\mathcal{N}_\ell; \ a \mapsto (\lambda_a, 1, \lambda_a)$$

is an isomorphism. In particular, $\mathcal{N}_\ell$ is a group.

(3) *The map $\mathcal{N}_\ell \to \mathcal{M}_\ell; \ a \mapsto \lambda_a$ is a monomorphism.*

Proof. (1) All statements are rewordings of the definition.

(2) $\mathrm{T}\mathcal{N}_\ell$ is the kernel of a homomorphism, hence a normal subgroup of $\mathrm{Top}(L)$.

An element of $\mathrm{T}\mathcal{N}_\ell$ is of the shape $(\alpha, 1, \gamma) \in \mathrm{Top}(L)$. By (4.2.1), it follows that $\alpha = \gamma = \lambda_a$, for $a = \alpha(1)$. Now, (1) implies that

τ_ℓ is a well-defined epimorphism. τ_ℓ is injective, because the map $a \mapsto \lambda_a$ is injective for every left loop.

(3) Indeed, the given map is just $\pi_1 \tau_\ell$, which is clearly injective. ∎

Only slightly more involved is

(5.2) *Let L be a loop, and let $T\mathcal{N}_m$ be the kernel of $\pi_3 : \mathrm{Top}(L) \to \mathcal{S}_L$. We have*

(1) *For $b \in L$, the following are equivalent*

   (I) $b \in \mathcal{N}_m$;

   (II) $(\varrho_b^{-1}, \lambda_b, 1) \in \mathrm{Top}(L)$;

  (III) $\lambda_x \lambda_b = \lambda_{xb}$ *for all $x \in L$*;

  (IV) $\delta_{x,b} = 1$ *for all $x \in L$*;

   (V) $\varrho_y \varrho_b = \varrho_{by}$ *for all $y \in L$*.

(2) *$T\mathcal{N}_m$ is a normal subgroup of $\mathrm{Top}(L)$, and the map*

$$\tau_m : \mathcal{N}_m \to T\mathcal{N}_m; \quad b \mapsto (\varrho_b^{-1}, \lambda_b, 1)$$

is an isomorphism, so $\mathcal{N}_m$ is a group.

(3) *The map $\mathcal{N}_m \to \mathcal{M}_\ell; \ b \mapsto \lambda_b$ is a monomorphism, and so is the map $\mathcal{N}_m \to \mathcal{M}_r; \ b \mapsto \varrho_b^{-1}$.*

Proof. (1) Only "(I) $\Longleftrightarrow$ (II)" requires a line of reasoning:

$$\forall x, y \in L : x \cdot by = \varrho_b^{-1}(xb)\lambda_b(y) = xb \cdot y$$

$$\Longleftrightarrow (\varrho_b^{-1}, \lambda_b, 1) \in \mathrm{Top}(L).$$

Note that xb runs through all of L as x runs through all of L.

(2) An element of $T\mathcal{N}_m$ has the form $(\alpha, \beta, 1) \in \mathrm{Top}(L)$. By (4.2.1)

$$\alpha = \varrho_b^{-1}, \ \beta = \lambda_a^{-1}, \quad \text{where } b = \beta(1), \text{ and } a = \alpha(1).$$

For all $y \in L$ we have

$$y = \varrho_b^{-1}(b)\lambda_a^{-1}(ay) = b \cdot ay \quad \text{hence} \quad \lambda_b \lambda_a = 1 \quad \text{and} \quad \beta = \lambda_b.$$

Now the argument can be concluded as in (5.1).

(3) The given maps are just $\pi_2 \tau_m$ and $\pi_1 \tau_m$, respectively. Both are clearly injective. ∎

Directly from the definition we obtain

(5.3) *Let L be a left loop, and $b \in L$. Then $b \in \mathcal{N}_r$ if and only if $\delta_{x,y}(b) = b$ for all $x, y \in L$. In particular $\mathcal{N}_r = \mathrm{Fix}\big(\mathcal{D}(L)\big)$.* ∎

Duality gives the theorem analogous to (5.1) for the right nucleus.

(5.4) *Let L be a loop, and let $\mathrm{T}\mathcal{N}_r$ be the kernel of $\pi_1 : \mathrm{Top}(L) \to \mathcal{S}_L$. We have*

(1) *For $b \in L$, the following are equivalent*

   (I) $b \in \mathcal{N}_r$;

  (II) $(1, \varrho_b, \varrho_b) \in \mathrm{Top}(L)$;

 (III) $\varrho_b \varrho_x = \varrho_{xb}$ for all $x \in L$;

 (IV) $\delta_{x,y}(b) = b$ for all $x, y \in L$.

(2) *$\mathrm{T}\mathcal{N}_r$ is a normal subgroup of $\mathrm{Top}(L)$, and the map*

$$\tau_r : \mathcal{N}_r \to \mathrm{T}\mathcal{N}_r; \; b \mapsto (1, \varrho_b^{-1}, \varrho_b^{-1})$$

is an isomorphism. Thus, $\mathcal{N}_r$ is a group.

(3) *The map $\mathcal{N}_r \to \mathcal{M}_r; \; b \mapsto \varrho_b^{-1}$ is a monomorphism.* ∎

We record an obvious consequence from (5.1), (5.2), and (4.3).

(5.5) *For every loop L we have $\mathrm{T}\mathcal{N}_\ell \cap \mathrm{T}\mathcal{N}_m = \{1\}$, hence $\mathrm{T}\mathcal{N}_\ell \times \mathrm{T}\mathcal{N}_m$ is a normal subgroup of $\mathrm{Top}(L)$.* ∎

We shall now give a proof for the fact that $\mathcal{N}_\ell = \mathcal{N}_m$ for left inverse property loops. This seems to have been proved first by Artzy [4; Cor. 2]. As a preparation we give a simple lemma, part (2) of which is due to Kinyon [69].

(5.6) *Let L be a left loop with unique inverses, then we have for all $a \in \mathcal{N}_m$, $x \in L$,*

(1) *$(xa)^{-1} = a^{-1}x^{-1}$ and $(ax)^{-1} = x^{-1}a^{-1}$.*

(2) *If L satisfies the automorphic inverse property, then $ax = xa$.*

Proof. (1) Noting that $a^{-1} \in \mathcal{N}_m$ by (5.2.2), we have

$$xa \cdot a^{-1}x^{-1} = x(a \cdot a^{-1}x^{-1}) = xx^{-1} = 1.$$

Similarly for the second assertion.

(2) By (1) $ax = (x^{-1}a^{-1})^{-1} = (x^{-1})^{-1}(a^{-1})^{-1} = xa$. ∎

We use isotopy theory for the proof of the main result.

(5.7) Theorem. *Let L be a left loop with left inverse property, and let $\iota : L \to L$; $x \mapsto x^{-1}$.*

(1) *If (α, β, γ) is an autotopism of L, then so is $(\iota\alpha\iota, \gamma, \beta)$.*

(2) $\mathcal{N}_\ell = \mathcal{N}_m$.

Proof. (1) For $x, y \in L$ we have

$$\gamma(y) = \gamma(x^{-1} \cdot xy) = \alpha(x^{-1})\beta(xy) \implies \alpha(x^{-1})^{-1}\gamma(y) = \beta(xy),$$

hence the result.

(2) Let $a \in \mathcal{N}_\ell$, then $(\lambda_a, 1, \lambda_a)$ is an autotopism, by (5.1.1). By (1) so is $(\iota\lambda_a\iota, \lambda_a, 1)$, which is in T$\mathcal{N}_m$. By (4.2.1) ϱ_a is invertible and $\iota\lambda_a\iota = \varrho_a^{-1}$. As in the proof of (5.2.1) one derives $a \in \mathcal{N}_m$.

Now let $a \in \mathcal{N}_m$, and pick $b, c \in L$. By (5.6.1)

$$(ba)^{-1} = a^{-1}b^{-1}.$$

Since L is a left loop with unique inverses there exists $b_1 \in L$ such that $a^{-1}b_1^{-1} = b$. Thus, we can compute

$$a \cdot bc = a \cdot (b_1 a)^{-1}c = b_1^{-1}\Big(b_1\big(a \cdot (b_1 a)^{-1}c\big)\Big)$$
$$= b_1^{-1}\big(b_1 a \cdot (b_1 a)^{-1}c\big) = b_1^{-1}c = (a \cdot a^{-1}b_1^{-1})c = ab \cdot c,$$

So $a \in \mathcal{N}_\ell$. ∎

Remarks. 1. If L was a loop, then the proof of $\mathcal{N}_m \subseteq \mathcal{N}_\ell$ could be done using isotopy theory, as well. Indeed, one simply applies (5.2.1) and (1), to conclude the proof as for the inclusion $\mathcal{N}_\ell \subseteq \mathcal{N}_m$.

2. Our requirement that every groupoid has an identity, makes sure that the nuclei are not empty. If we had a quasigroup L to begin with, then a nucleus (and hence every nucleus) is non-empty if and only if L is a loop (see [94; I.3.4]).

The *center* of a groupoid is defined by

$$\mathcal{Z}(L) := \{a \in \mathcal{N}_\ell \cap \mathcal{N}_m \cap \mathcal{N}_r; \ \forall x \in L : ax = xa\}.$$

For simplicity, the definition has been kept symmetric. In fact, it suffices to take elements from the intersection of any two of the nuclei to define the center. We make this explicit in one case.

(5.8) *Let L be a groupoid, then $\mathcal{Z}(L) = \{a \in \mathcal{N}_\ell \cap \mathcal{N}_m;\ \forall x \in L : ax = xa\}$.*

Proof. It suffices to show that $a \in \mathcal{N}_\ell \cap \mathcal{N}_m$ with $ax = xa$ for all $x \in L$ is an element of $\mathcal{N}_r$. Indeed, we can compute for all $x, y \in L$

$$xy \cdot a = a \cdot xy = ax \cdot y = xa \cdot y = x \cdot ay = x \cdot ya,$$

thus $a \in \mathcal{N}_r$. ∎

It should be emphasized that instead of $\mathcal{N}_\ell \cap \mathcal{N}_m$ also $\mathcal{N}_\ell \cap \mathcal{N}_r$ and $\mathcal{N}_m \cap \mathcal{N}_r$ qualify, with very similar proofs.

(5.9) *Let a be an element of a left loop L.*

(1) *If $a \in \mathcal{Z}(L)$, then λ_a centralizes $\mathcal{M}_\ell$.*

(2) *If λ_a centralizes $\mathcal{D}(L)$, then $a \in \mathcal{N}_r$.*

(3) *If L is A_ℓ, then λ_a centralizes $\mathcal{D}(L)$ if and only if $a \in \mathcal{N}_r$.*

Proof. (1) $\lambda_a \lambda_x = \lambda_{ax} = \lambda_{xa} = \lambda_x \lambda_a$ for all $x \in L$.

(2) For all $x, y \in L$ we have $\lambda_a \delta_{x,y} = \delta_{x,y} \lambda_a$. Apply this to 1 to see $\delta_{x,y}(a) = a$, hence $a \in \mathcal{N}_r$ by (5.3).

(3) Let $a \in \mathcal{N}_r$. For $\delta \in \mathcal{D}(L)$ using (5.3) we have $\delta(ax) = \delta(a)\delta(x) = a\delta(x)$. Hence $\delta \lambda_a = \lambda_a \delta$. ∎

Remarks. 1. Assume λ_a centralizes $\mathcal{M}_\ell$, then clearly $ax = xa$ for all $x \in L$. However, it can be shown by example that a is not necessarily in $\mathcal{Z}(L)$. Indeed, there exists a Bol loop L of order 16 with trivial center such that $\lambda(L) \cap \mathcal{Z}(\mathcal{M}_\ell) \neq \{\mathbf{1}\}$. It has been generated using GAP [39].

2. This phenomenon does not occur in the full multiplication group $\mathcal{M}$. Then λ_a (as well as ϱ_a) centralizes $\mathcal{M}$ if and only if $a \in \mathcal{Z}(L)$, see [1; Thm. 11]. This has the consequence that isotopic loops have isomorphic centers, [1; Thm. 12].

3. From (1), (2), and (2.3) it is easy to see that L is a commutative group if and only if $\mathcal{M}_\ell(L)$ is a commutative group (see also [68; Lemma 2]).

The following applies to Kikkawa left loops, and in particular to K-loops, as we shall see in (6.7).

(5.10) Theorem. *Let L be a left loop with left and automorphic inverse property, then*

(1) $\mathcal{Z}(L) = \mathcal{N}_\ell = \mathcal{N}_m \subseteq \mathcal{N}_r.$

(2) *If $\mathcal{D}(L)$ acts fixed point free on L, then $\mathcal{Z}(L) = \mathcal{N}_\ell = \mathcal{N}_m = \mathcal{N}_r = \{1\}$, or $L = \mathcal{Z}(L)$ is an abelian group.*

Proof. (1) is direct from (5.7.2), (5.8), and (5.6.2).

(2) follows directly from (5.3) and (1), since either $\mathcal{D}(L) = \{1\}$ or $\mathrm{Fix}(\mathcal{D}(L)) = \{1\}$ by hypothesis.   ∎

Remark. Part (1) of the previous theorem can be derived from the identity $a(ab)^{-1} = b^{-1}$ alone (cf. (3.5)). This is due to JOHNSON and SHARMA [51; Thm. 6].

6. Bol Loops and K-Loops

Before we continue the discussion of Bol loops, we give a characterization of left power alternativity. Recall that a groupoid L is
called *left power alternative* if every element $a \in L$ has a unique
inverse and if $\lambda_a^k = \lambda_{a^k}$, $\forall k \in \mathbf{Z}$. According to (3.1.1) L satisfies
the left inverse property and is therefore a left loop. It will turn
out soon that Bol loops are left power alternative.

A. LEFT POWER ALTERNATIVE LEFT LOOPS

The notion of left power alternativity and the basic content of the
following lemma (but with stronger hypothesis) can be found in
[67; Prop. 1.11].

(6.1) *Let L be a left loop, and $a \in L$ with unique inverse a^{-1}.*

(1) *The following are equivalent*

  (I) *a is left power alternative;*

  (II) $\forall k, \ell \in \mathbf{Z}, x \in L : a^k \cdot a^\ell x = a^{k+\ell} x$;

  (III) $\forall k, \ell \in \mathbf{Z} : \delta_{a^k, a^\ell} = 1$;

  (IV) $\forall k \in \mathbf{Z} : \delta_{a, a^k} = 1$;

  (V) $\forall k \in \mathbf{Z} : \delta_{a^k, a} = 1$, *and* $\delta_{a^k, a^{-1}} = 1$.

(2) *If the conditions in (1) are satisfied, then a is contained in a
cyclic subgroup of L, a is alternative and satisfies the left inverse
property. Furthermore, it makes sense to write $\langle a \rangle$ for the subgroup
generated by a, and $\langle a \rangle \cong \langle \lambda_a \rangle$.*

(3) *If moreover, L is a finite loop, then the order[1] $|a|$ of a divides
$|L|$, the order of L.*

Proof. (1) The following is easy (see also (3.2))

$$\lambda_{a^{-1}} \lambda_a = 1 \iff \lambda_a \lambda_{a^{-1}} = 1 \iff \delta_{a^{-1}, a} = 1 \iff \delta_{a, a^{-1}} = 1.$$

---

[1] Since a is contained in a subgroup, this notion can be taken from
group theory.

This characterizes a satisfying the left inverse property.

(I) $\Longrightarrow$ (II): For all $x \in L$, $k, \ell \in \mathbf{Z}$, we have

$$a^k \cdot a^\ell x = \lambda_{a^k} \lambda_{a^\ell}(x) = \lambda_a^k \lambda_a^\ell(x) = \lambda_a^{k+\ell}(x) = \lambda_{a^{k+\ell}}(x) = a^{k+\ell}x.$$

(II) $\Longrightarrow$ (III): Let $k, \ell \in \mathbf{Z}$. Then for all $x \in L$ we compute $a^k \cdot a^\ell x = a^{k+\ell} x = a^k a^\ell \cdot x$. This implies $\delta_{a^k, a^\ell} = 1$.

"(III) $\Longrightarrow$ (IV)" and "(III) $\Longrightarrow$ (V)" are trivial.

(IV) $\Longrightarrow$ (I): We have to prove that $\lambda_{a^k} = \lambda_a^k$ for all $k \in \mathbf{Z}$. The case $k \geq 0$ is done by induction. Using $\delta_{a, a^k} = 1$, and the induction hypothesis, we compute

$$\lambda_{a^{k+1}} = \lambda_{aa^k} = \lambda_a \lambda_{a^k} \delta_{a, a^k}^{-1} = \lambda_a \lambda_a^k = \lambda_a^{k+1}.$$

Using $k = -1$, the remark at the beginning of the proof implies the left inverse property for a. Therefore, the case $k < 0$ can be proved similarly using $\delta_{a^{-1}, a^\ell} = \delta_{a, a^{\ell-1}}^{-1} = 1$, $\forall \ell \in \mathbf{Z}$, from (3.2).

(V) $\Longrightarrow$ (I): Assume $k > 0$. We first prove by induction on $k \in \mathbf{N}$ that $a^k a^\ell = a^{k+\ell}$ for all $\ell \in \mathbf{N}$. Notice that the induction base, $k = 1$, holds by the definition of powers.

$$a^k a^\ell = a^{k-1} a \cdot a^\ell = a^{k-1} \cdot a \delta_{a^{k-1}, a}^{-1}(a^\ell) = a^{k-1} \cdot a^{\ell+1} = a^{k+\ell}.$$

Now we can do induction on $k \in \mathbf{N}$ again to obtain

$$\lambda_{a^{k+1}} = \lambda_{a^k a} = \lambda_{a^k} \lambda_a \delta_{a^k, a}^{-1} = \lambda_a^k \lambda_a = \lambda_a^{k+1}.$$

Exactly the same line of reasoning can be used to do the case $k < 0$. Therefore, a is left power alternative.

(2) Recall that the left inverse property has been taken care of. The case $k = 1$ in (IV) shows that a is alternative.

Furthermore, observe that by (II) the map $\mathbf{Z} \to L; k \mapsto a^k$ is a homomorphism. Therefore, the set $A := \{a^k; k \in \mathbf{Z}\}$ is a (necessarily cyclic) group, which clearly contains a.

The map $\langle a \rangle \to \langle \lambda_a \rangle$; $a^k \mapsto \lambda_a^k$ is an isomorphism.

(3) We'll show that the right cosets of $\langle a \rangle$ form a partition of L. Let $b, c \in L$ such that $\langle a \rangle b \cap \langle a \rangle c \neq \varnothing$, i.e., there are $n, m \in \mathbf{Z}$ with $a^n b = a^m c$. Using (II), we can compute $a^\ell b = a^{m-n+\ell} c \in \langle a \rangle c$ for every $\ell \in \mathbf{Z}$. Hence $\langle a \rangle b \subseteq \langle a \rangle c$. By symmetry, we must have $\langle a \rangle b = \langle a \rangle c$. Since L is a loop, we have $|\langle a \rangle| = |\langle a \rangle b|$. This implies $|a| = |\langle a \rangle|$ divides $|L|$. ∎

As indicated in the footnote to the previous theorem, we can use the notion of the *order* of an element in a left power alternative left loop in a reasonable manner. Likewise, it makes sense to speak of the *exponent* of such a loop (see §1.A).

The displayed chain of equations in the proof of "(I) $\Longrightarrow$ (II)" can be copied word by word to show

(6.2) *Let L be a groupoid, $a \in L$, and $n \in \mathbf{N}$. Assume that $\lambda_{a^j} = \lambda_a^j$ for all $j \in \mathbf{N}$, $j \leq n$, then for $k, \ell \in \mathbf{N}$, such that $k + \ell \leq n$ we have $a^k \cdot a^\ell x = a^{k+\ell} x$ for all $x \in L$.* ∎

This will be convenient for induction proofs later.

(6.3) *Let L be a left loop with left inverse property. If $\delta_{a,a^k} = 1$ for all $a \in L$, $k \in \mathbf{N}$, then L is left power alternative.*

Proof. Let $a \in L$. In view of (6.1.1)(IV), all we need to show is that $\delta_{a,a^{-k}} = 1$ for all $k \in \mathbf{N}$. Using the left inverse property we can compute

$$aa^{-k} = (a^{-1})^{-1} \cdot a^{-1}(a^{-1})^{k-1} = (a^{-1})^{k-1}$$

thus by (3.2) $\quad \delta_{a,a^{-k}} = \left(\delta_{a^{-1},(a^{-1})^{k-1}}\right)^{-1} = 1.$ ∎

Remarks. 1. The hypothesis "L is a loop" in (6.1.3) is necessary. If L is only a left loop, then (12.3) provides counterexamples. Still, the right cosets of $\langle a \rangle$ form a partition of L, but their sizes may vary.

2. If in a left loop L every element is contained in a cyclic subgroup as in (6.1.2), then L is called *power associative*. This is an important concept used a lot in loop theory. However, it does not suit out needs here, because all relevant structures satisfy the stronger condition of left power alternativity. The main advantage is that $|\lambda_a| = |a|$ for all $a \in L$. Also, (6.1.3) does not in general follow for power associative loops.

B. BOL LOOPS

The existence of certain autotopisms, respectively isotopes, give a way to characterize Bol loops. We give a streamlined exposition of results in [100] and [95], supplemented by some identities for left inner mappings and some well-known trivialities.

(6.4) Theorem. *Let L be a loop.*

(1) *The following are equivalent*

   (I) L *is Bol, i.e.,* $\forall a, b, c \in L : a(b \cdot ac) = (a \cdot ba)c$;

  (II) $\lambda_a \lambda(L) \lambda_a \subseteq \lambda(L)$ *for all* $a \in L$;

 (III) $\lambda_a \lambda_b \lambda_a = \lambda_{a \cdot ba}$ *for all* $a, b \in L$;

 (IV) $\delta_{ab,b} = \delta_{a,b}$ *for all* $a, b \in L$;

  (V) $\delta_{a,ba} = \delta_{b,a}^{-1}$ *for all* $a, b \in L$;

 (VI) $\delta_{a,ba} = \delta_{a,b}$ *and* $\delta_{a,b} = \delta_{b,a}^{-1}$ *for all* $a, b \in L$;

 (VII) $\tau_a := (\lambda_a \varrho_a, \lambda_a^{-1}, \lambda_a) \in \mathrm{Top}(L)$ *for every* $a \in L$;

(VIII) *Every isotope of L satisfies the left inverse property;*

 (IX) *Every isotope of L is left alternative;*

  (X) $\lambda_a \varrho_{ab} = \varrho_b \lambda_a \varrho_a$ *for all* $a, b \in L$.

(2) *Every isotope of a Bol loop is Bol.*

(3) *Bol loops are left power alternative.*

(4) *If L is a finite Bol loop, then the order of an element in L divides the order of L.*

Proof. (1) We follow the scheme:

$$(\mathrm{VIII}) \to (\mathrm{VII})$$

$$(\mathrm{X}) \leftrightarrow (\mathrm{I}) \qquad\qquad\qquad (\mathrm{V}) \leftarrow (\mathrm{VI})$$

$$(\mathrm{IX}) \to (\mathrm{II}) \leftarrow (\mathrm{III}) \leftarrow (\mathrm{IV})$$

For a preliminary remark, assume that L satisfies the left inverse property. Let $a, b, x, y \in L$ and put $x \circ y := \varrho_b^{-1}(x) \lambda_a^{-1}(y)$. Then $L^{(a,b)} = (L, \circ)$ is a principal isotope of L. We'll first derive a condition for $(L, \circ)$ to satisfy the left inverse property, as well. For

$x' \in L$ with $xb \circ x' = ab$, we have

$$x' = a(x^{-1} \cdot ab).$$

Note that ab is the identity element of $(L, \circ)$, so x' is the inverse of xb. Hence the left inverse property $x' \circ (xb \circ ay) = ay$ for $(L, \circ)$ is equivalent to

$$\varrho_b^{-1}\big(a(x^{-1} \cdot ab)\big)\lambda_a^{-1}(xy) = ay = \lambda_a(y). \qquad (i)$$

(I) $\implies$ (VIII): By (3.10), L satisfies the left inverse property, and the previous remark applies. Noting that $\varrho_b^{-1}\big(a(x^{-1} \cdot ab)\big) = \varrho_b^{-1}\big((a \cdot x^{-1}a)b\big) = a \cdot x^{-1}a$ (by Bol), we can compute

$$\begin{aligned}
\varrho_b^{-1}\big(a(x^{-1} \cdot ab)\big)\lambda_a^{-1}(xy) &= (a \cdot x^{-1}a)\lambda_a^{-1}(xy) \\
&= a\big(x^{-1} \cdot a\lambda_a^{-1}(xy)\big) \\
&= a(x^{-1} \cdot xy) = ay.
\end{aligned}$$

Thus the principal isotope $L^{(a,b)} = (L, \circ)$ satisfies the left inverse property. Since a, b are arbitrary, the result follows from (4.2.2).

(VIII) $\implies$ (VII): By assumption, the preliminary remark applies here, too, and (i) holds. Choosing $b = 1$ gives

$$\lambda_a\varrho_a(x^{-1})\lambda_a^{-1}(xy) = \lambda_a(y) = \lambda_a(x^{-1} \cdot xy), \quad \text{for all } x, y \in L.$$

hence $(\lambda_a\varrho_a, \lambda_a^{-1}, \lambda_a) \in \mathrm{Top}(L)$.

(VII) $\implies$ (V): For $a, b, x \in L$ we have

$$\begin{aligned}
a\lambda_b(x) = \lambda_a(bx) = \lambda_a\varrho_a(b)\lambda_a^{-1}(x) &= (a \cdot ba)\lambda_a^{-1}(x) \\
&= a\big(ba \cdot \delta_{a,ba}^{-1}\lambda_a^{-1}(x)\big).
\end{aligned}$$

After canceling a from the leftmost and rightmost expression, we obtain

$$\lambda_b = \lambda_{ba}\delta_{a,ba}^{-1}\lambda_a^{-1}, \quad \text{hence} \quad \delta_{a,ba}^{-1} = \lambda_{ba}^{-1}\lambda_b\lambda_a = \delta_{b,a}.$$

(V) $\implies$ (IV): For $a \in L$ let $a' \in L$ be such that $a'a = 1$, then $\delta_{a',a} = \delta_{a,a'a}^{-1} = \delta_{a,1}^{-1} = 1$. By (3.1.1) L satisfies the left inverse property. Using this and (3.2) we can compute

$$\delta_{a,b} = \delta_{a^{-1},ab}^{-1} = \delta_{ab,a^{-1}\cdot ab} = \delta_{ab,b} \quad \text{for all } a,b \in L.$$

(IV) $\implies$ (III): For $a \in L$ let $a' \in L$ be such that $a'a = 1$, then $\delta_{a',a} = \delta_{a'a,a} = \delta_{1,a} = 1$. By (3.1.1) again, L satisfies the left inverse property. This gives for all $a,b \in L$

$$\delta_{a,ba} = \delta_{b^{-1}\cdot ba,ba} = \delta_{b^{-1},ba},$$

and using the left inverse property

$$\lambda_{a\cdot ba}^{-1}\lambda_a\lambda_{ba} = \lambda_a^{-1}\lambda_{b^{-1}}\lambda_{ba} = \lambda_a^{-1}\lambda_b^{-1}\lambda_{ba}.$$

Canceling λ_{ba} and rearranging terms shows the result.

"(III) $\implies$ (II)" and "(II) $\implies$ (I)" are in (3.9), "(VI) $\implies$ (V)" and "(I) $\iff$ (X)" are trivial.

This shows the equivalence of (I)–(V), (VII), (VIII). From these we derive (VI) and the supplementary statements.

(IV),(V) $\implies$ (VI): For $a,b \in L$, let $c \in L$ with $cb = a$. Then (IV) and (V) show

$$\delta_{a,b} = \delta_{cb,b} = \delta_{c,b} = \delta_{b,cb}^{-1} = \delta_{b,a}^{-1}, \quad \text{the second identity.}$$

The first identity is now obvious.

(2) is a consequence of "(I) $\iff$ (VIII)", and (4) follows directly from (3) and (6.1.3).

(3) We first prove by induction on k that $\lambda_a^k = \lambda_{a^k}$ for all $k \in \mathbf{N}$. Using (III), (6.2) together with the induction hypothesis, and the definition of powers in a loop (in that order), we find

$$\lambda_a^k = \lambda_a\lambda_a^{k-2}\lambda_a = \lambda_{a\cdot a^{k-2}a} = \lambda_{a\cdot a^{k-1}} = \lambda_{a^k}.$$

Now the case $k \leq 0$ easily follows using the left inverse property (see (3.10)).

"(I) $\implies$ (IX)" is clear from (2) and (3).

(IX) $\implies$ (II): For $a' \in L$ take $a \in L$ with $aa' = 1$. The fact that the principal isotope $(L, \circ) = L^{(a,1)}$ is alternative implies for all $x, y \in L$

$$x \circ (x \circ ay) = (x \circ x) \circ ay, \quad \text{thus}$$

$$x\lambda_a^{-1}(xy) = x\lambda_a^{-1}(x) \cdot y = \lambda_{x\lambda_a^{-1}(x)}(y).$$

If we put $x = 1$, then $\lambda_a^{-1}(y) = \lambda_{a'}(y)$, since $a' := \lambda_a^{-1}(1)$. Therefore $\lambda_x \lambda_{a'} \lambda_x \in \lambda(L)$ for all $x \in L$. ∎

Remarks. 1. (III) and (X) are the only possibilities to express the Bol identity using just the multiplication maps. (X) appears in [103].

2. (IV) was used by UNGAR [116, 117], and later called "loop property" in his axioms for "gyrogroups" (see e.g. [118]). The equivalence with Bol was proved in [108]. Thus gyrogroups are Bol loops. (V) appeared in $[81; (2.11)]^2$. The equivalence of (VI) and Bol is due to KINYON [71].

3. "(I) $\iff$ (VIII)" seems to have appeared in [95; p. 57] first.[3] Other proofs are in [100; Thm. 3.1] and [10; 10.4.6, p. 191]. "(I) $\iff$ (IX)" is also in [100; Thm. 3.1]. Condition (VII) shows up in [100; Thm. 2.3].

4. A proof of (2), independent of (1), is given in [10; Thm. 6.10, p. 106]. ROBINSON [100] uses our approach.

5. Part (4) of the preceding theorem is a special case of LAGRANGE's theorem for finite Bol loops. This has been first proved by BURN [22; Cor. 1], but his proof is valid in every left power alternative loop. A weaker version occurs in [20; V.1.2, p. 92].

Up to isomorphism, the isotopes of Bol loops can be taken rather special. See [94; IV.6.16, p. 119] for a proof.

(6.5) Theorem. *Let L be a Bol loop, $a, b \in L$ and let $L^{(a,b)}$ be a principal isotope. Then $L^{(a,b)}$ is isomorphic to $L^{(c^{-1}, c)}$, where $c = ab \cdot b$. Therefore, every isotope of L is isomorphic to a principal isotope $L^{(c^{-1}, c)}$ for suitable $c \in L$.* ∎

---

[2] In view of (3.11), this is only formally more general than our (V).

[3] The first edition of PICKERT's book was published in 1955.

C. K-Loops

A Bol loop which satisfies the automorphic inverse property is called a *K-loop* or *Bruck loop*. The next theorem is extremely important for the axiomatics of K-loops.

(6.6) Theorem. *Let L be a Bol loop and $a, b \in L$.*

(1) $\delta_{a,b} \in \Psi(L)$ *is a pseudoautomorphism with companion* $ab \cdot a^{-1}b^{-1}$.

(2) $(ab)^{-1} = a^{-1}b^{-1} \implies \delta_{a,b} \in \operatorname{Aut} L$.

(3) *If L satisfies the automorphic inverse property, i.e., if L is a K-loop, then L is an A_ℓ-loop.*

Proof. (1) Let $\tau_a = (\lambda_a \varrho_a, \lambda_a^{-1}, \lambda_a)$. By (6.4.1)(VII),

$$\tau_{ab}\tau_a^{-1}\tau_b^{-1} = (\,.\,, \delta_{a,b}, \lambda_{ab}\lambda_a^{-1}\lambda_b^{-1})$$

is an autotopism. Since $\delta_{a,b}(1) = 1$, we get from (4.4.3) that $\delta_{a,b} \in \Psi(L)$, with companion $\lambda_{ab}\lambda_{a^{-1}}\lambda_{b^{-1}}(1) = ab \cdot a^{-1}b^{-1}$. The third component of the above autotopism has been rewritten, using the left inverse property.

(2) and (3) are direct consequences of (1). ∎

Remarks. 1. Part (3) of this theorem has been proved in three papers recently. FUNK and NAGY in [34; 5.1] use a geometric argument in the corresponding net. KREUZER [80] gives a direct proof, avoiding the use of pseudoautomorphisms. We have adopted the more general [41; 3.12] by GOODAIRE and ROBINSON, which is our (1). This in turn generalizes [20; VII Lemma 2.2, p. 113] for Moufang loops.

2. There exist (even finite) Bol loops L such that $\delta_{a,b}$ is not an automorphism for some $a, b \in L$. Indeed, every simple non-associative Moufang loop L has this property, because there exist $a, b \in L$ with $u := aba^{-1}b^{-1} \neq 1$. Thus $u \notin \mathcal{N}_\ell(L) = \{1\}$ and using (4.4.1) one derives that $\delta_{a,b}$ is a proper pseudoautomorphism. PAIGE [92] has constructed an infinite class of finite simple non-associative Moufang loops.

We now give the theorem of KREUZER [80; 3.4], which shows that the formerly used definition of a K-loop (Kikkawa loop with $\delta_{a,b} = \delta_{a,ba}$)[4] is equivalent with ours.

(6.7) Theorem. *Let L be a groupoid. Then L is a K-loop if and only if L is a Kikkawa loop, and $\delta_{a,b} = \delta_{a,ba}$ for all $a, b \in L$.*

Proof. Notice, both K-loops and Kikkawa loops are automorphic inverse by definition.

Let L be a K-loop. By (3.10.3) and (6.6.3) L is a Kikkawa loop. (6.4.1)(VI) completes this direction.

For the converse, the hypothesis and (3.7.3) imply (6.4.1)(VI), so L is Bol. ∎

The implication "(I) $\Longrightarrow$ (III)" of the following is [40; Lemma 1]. GLAUBERMAN [40] attributes the converse to ROBINSON.

(6.8) *Let L be a Bol loop.*

(1) *The following are equivalent*

   (I) L *is a K-loop;*

  (II) $\lambda_{ab}^2 = \lambda_a \lambda_b^2 \lambda_a$ *for all $a, b \in L$;*

 (III) $(ab)^2 = a \cdot b^2 a$ *for all $a, b \in L$.*

(2) *If L is a K-loop, then the map $\kappa : x \mapsto x^2$ is injective if and only if L contains no elements of order 2.*

Proof. (1) In view of (3.10.3) and (6.6), (3.7.1) and (3.6.3) show that (I) and (II) are equivalent.

(II) $\Longrightarrow$ (III): Applying both sides to 1 and using the fact that Bol loops are left alternative gives the claimed identity.

(III) $\Longrightarrow$ (II): We use (6.4.1)(III) and (6.4.3) to compute

$$\lambda_{ab}^2 = \lambda_{(ab)^2} = \lambda_{a \cdot b^2 a} = \lambda_a \lambda_{b^2} \lambda_a = \lambda_a \lambda_b^2 \lambda_a.$$

(2) is a direct consequence of (3.8) and (III). ∎

If the squaring map behaves well, we get a much stronger result.

(6.9) Theorem. *Let L be a groupoid with right inverses.*

---

[4] This definition is directly derived from simple properties of neardomains, see [121; V §1].

(1) *For the following conditions, we have*
(I) $\Longrightarrow$ *(II)* $\Longrightarrow$ *(III)* $\Longrightarrow$ *(IV).*

(I)   *L is a K-loop;*

(II)   *L is a left alternative Kikkawa loop;*

(III)   *L is left alternative, satisfies the left inverse property, the automorphic inverse property, and A_ℓ;*

(IV) $\forall a, b \in L : \lambda_{ab}^2 = \lambda_a \lambda_{b^2} \lambda_a.$

(2) *If the map $L \to L$; $x \mapsto x^2$ is surjective, then the preceding conditions are equivalent.*

Proof. (1) "(I) $\Longrightarrow$ (II)" is direct from (6.4.3) and (6.7), "(II) $\Longrightarrow$ (III)" is trivial.

(III) $\Longrightarrow$ (IV): By (3.1), L is a left loop, hence (3.7.3) is applicable, and yields together with the left alternative property

$$\lambda_{ab}^2 = \lambda_a \lambda_b^2 \lambda_a = \lambda_a \lambda_{b^2} \lambda_a \,.$$

(2) (IV) $\Longrightarrow$ (I): Putting $a = 1$, we obtain $\lambda_b^2 = \lambda_{b^2}$ for all $b \in L$, i.e., L is left alternative. Let now $a, b \in L$. By assumption there exists $d \in L$ with $d^2 = b$. So we can compute

$$\lambda_a \lambda_b \lambda_a = \lambda_a \lambda_{d^2} \lambda_a = \lambda_{ad}^2 = \lambda_{(ad)^2} \in \lambda(L).$$

In view of (3.9) and (3.10), we conclude that L is a Bol loop and satisfies the left inverse property. Since L is left alternative, $\lambda_{ab}^2 = \lambda_a \lambda_{b^2} \lambda_a = \lambda_a \lambda_b^2 \lambda_a$. Now we can invoke (3.6.3) or (6.8.1) to conclude that L satisfies the automorphic inverse property.   ∎

Remarks. 1. This theorem and its proof have been compiled from [88; XII.3.29, 3.34, 3.35]. The topological hypotheses used there imply 2-divisibility. This is all that's really needed for (2).

2. (IV) is equivalent to either of the following

$$\forall a, b \in L : \lambda_{(ab)^2} = \lambda_a \lambda_{b^2} \lambda_a \quad \text{or} \quad \forall a, b \in L : \lambda_{(ab)^2} = \lambda_a \lambda_b^2 \lambda_a \,.$$

Indeed, putting $b = 1$, or $a = 1$, respectively, it is seen that both conditions imply that L is left alternative. Then the claimed equivalence is obvious.

3. The identity $\lambda_{ab}^2 = \lambda_a \lambda_b^2 \lambda_a$ for all $a, b \in L$ from (6.8.1) is true in every Kikkawa loop (see (3.7.3)). By the example in (12.9.3) Kikkawa loops need not be left alternative. Therefore this identity is not equivalent to the identities of the previous remark. See also [73; Thm. 4.1].

4. KREUZER [79; (3.5)] shows by examples that the implication "(II) $\Longrightarrow$ (I)" is not true in general. Indeed, in (12.9.5) we present his construction of a left power alternative Kikkawa loop, which is not a Bol loop.

5. If L is a Lie loop, then the hypothesis in (II) that L be left alternative is redundant, see [67; Lemma 6.2]. Notice that from (3.7.3) one easily gets that $\delta_{a,a}^2 = 1$ for all $a \in L$. If L is connected, then it can also be seen that the square map is surjective. Hence KIKKAWA's connected, symmetric Lie loops are in fact K-loops (see also [88; XII.3.34]).

A left power alternative left loop L will be called *(uniquely) n-divisible*, $n \in \mathbf{N}$, if for every $a \in L$ there exists (exactly one) $b \in L$ with $b^n = a$. Notice that by (6.1.2) a, b are contained in an abelian (in fact cyclic) subgroup of L, so this notion coincides with the standard definition [33; §20]. In particular, we can write $b = a^{\frac{1}{n}}$ if L is uniquely n-divisible. In this case it also makes sense to write $a^{\frac{k}{n}}$ for every $k \in \mathbf{Z}$. Abusing language, we shall say that "$a^{\frac{k}{n}}$ is well-defined". Finally, note that a loop is n-divisible if and only if the map $x \mapsto x^n$ is surjective, it is uniquely n-divisible if and only if the map $x \mapsto x^n$ is bijective.

(6.10) *Let L be a left power alternative A_ℓ-left-loop such that for a fixed rational number q the map $L \to L$; $x \mapsto x^q$ is well-defined. Then $a(a^{-1}b)^q = b(b^{-1}a)^{1-q}$ for all $a, b \in L$.*

Proof. Using (3.1.1) and (6.1.1), we can compute

$$b^{-1} \cdot a(a^{-1}b)^q = b^{-1}a \cdot \delta_{b^{-1},a}(a^{-1}b)^q = b^{-1}a \cdot (b^{-1}a)^{-q}$$
$$= (b^{-1}a)^{1-q}.$$

Multiplying with b on both sides gives the result. ∎

Remark. The proof only uses that the maps $x \mapsto x^q$ and $\delta_{b^{-1},a}$ commute. A_ℓ is not really needed.

(6.11) Theorem. *Let L be a uniquely 2-divisible left power al-ternative A_ℓ-left-loop, and let ε be an involutory, fixed point free automorphism of L. Then $\varepsilon(x) = x^{-1}$ for all $x \in L$, and L is a K-loop.*

Proof. By (6.10) we find

$$\varepsilon\left(x\left(x^{-1}\varepsilon(x)\right)^{\frac{1}{2}}\right) = \varepsilon(x)\left(\varepsilon(x)^{-1}x\right)^{\frac{1}{2}} = x\left(x^{-1}\varepsilon(x)\right)^{1-\frac{1}{2}}$$

$$= x\left(x^{-1}\varepsilon(x)\right)^{\frac{1}{2}}.$$

Therefore the assumption implies $x\left(x^{-1}\varepsilon(x)\right)^{\frac{1}{2}} = 1$. Using the left inverse property we get $x^{-1}\varepsilon(x) = x^{-2}$, and then $\varepsilon(x) = x^{-1}$. This shows that L has the automorphic inverse property. Since 2-divisible simply means that the square map $x \mapsto x^2$ is surjective, we can conclude from (6.9) that L is a K-loop. ∎

Remark. The theorem and its proof, including the preceding lemma, are due to KIST [74; (1.2.e), (1.4.b)]. We have only modi-fied the context to obtain the presented generalization.

To conclude this subsection, we present without proof ROBINSON's results on the isotopes of K-loops [100].

(6.12) Theorem. *Let L be a K-loop and $c \in L$. The isotope $L^{(c^{-1},c)}$ is a K-loop if and only if $c \in N_r$.*

Proof. [100; Thm. 3.2]. ∎

As a corollary using (6.5) and the fact that loop isotopes of groups are groups, we get

(6.13) Theorem. *Let L be a K-loop. The following are equivalent*

   (I)   *Every isotope of L is a K-loop;*

   (II) *Every isotope of L is isomorphic to L, i.e., L is a G-loop;*

   (III) *L is an abelian group.* ∎

D. HALF EMBEDDING

In [40] GLAUBERMAN studied finite K-loops L of odd order. By (6.4.4), L contains no elements of even order. From (6.8.2) we infer that L is uniquely 2-divisible. The converse is also true, i.e., a finite uniquely 2-divisible K-loop is of odd order. This can be derived easily from [40; Cor. 1, p. 394]. Actually, GLAUBERMAN states his results for, what he calls, *B-loops*, i.e., K-loops such that every element has finite odd order. He remarks that some of these results "might also remain valid" for the more general uniquely 2-divisible K-loops.

The basic idea for the following construction is already in [20; VII, Thm. 5.2, p. 121]. It has been modified in [40; Lemma 3] with the hypotheses mentioned above. Our slightly more general approach seems to be used in [66; Thm. 5] first.

(6.14) Theorem. *Let G be a group with a subset L containing 1, i.e., $1 \in L \subseteq G$. Assume that $L^{-1} \subseteq L$, $aLa \subseteq L$ for all $a \in L$, and assume that the map $\kappa : L \to L$; $x \mapsto x^2$ is well-defined and bijective. For $a, b \in L$ write $a^{\frac{1}{2}} = \kappa^{-1}(a)$, and let*

$$a \circ b := a^{\frac{1}{2}} b a^{\frac{1}{2}},$$

then

(1) *$(L, \circ)$ is a uniquely 2-divisible K-loop with identity 1. Moreover, integer powers of elements in L formed in G and in $(L, \circ)$ coincide.*

(2) *L is commutative if and only if $[a, b] = [a^{-1}, b^{-1}]$ for all $a, b \in L$.*

(3) *L is a commutative group if and only if $[[a, b], c] = 1$ for all $a, b, c \in L$.*

Proof. (1) The assumptions imply directly that L is a groupoid with identity 1. Let $a, b, c \in L$. Clearly, $(a^{-1})^{\frac{1}{2}} = (a^{\frac{1}{2}})^{-1}$, therefore we can write $a^{-\frac{1}{2}}$ unambiguously. From $a^{-1} \circ (a \circ b) = a^{-\frac{1}{2}} a^{\frac{1}{2}} b a^{\frac{1}{2}} a^{-\frac{1}{2}} = b$, we find that L satisfies the left inverse property, hence L is a left loop by (3.1.1). Moreover, a^{-1} is also the inverse of a with respect to "$\circ$".

We have $(aa^{\frac{1}{2}}a^{-1})^2 = a$, hence $aa^{\frac{1}{2}}a^{-1} = a^{\frac{1}{2}}$, and $aa^{\frac{1}{2}} = a^{\frac{1}{2}}a$. By a simple induction one can deduce that integer powers of a formed in G or L are the same. They will therefore not be distinguished in our notation.

By assumption there exists $u \in L$ with $u = a^{\frac{1}{2}}b^{\frac{1}{2}}a^{\frac{1}{2}}$, so we can compute

$$a \circ (b \circ (a \circ c)) = a^{\frac{1}{2}}b^{\frac{1}{2}}a^{\frac{1}{2}}ca^{\frac{1}{2}}b^{\frac{1}{2}}a^{\frac{1}{2}} = ucu = u^2 \circ c.$$

Substituting $c = 1$ yields $u^2 = a \circ (b \circ a)$, hence L is Bol, and by (3.11) L is actually a Bol loop.

The calculation $a^{-1} \circ b^{-1} = a^{-\frac{1}{2}}b^{-1}a^{-\frac{1}{2}} = \left(a^{\frac{1}{2}}ba^{\frac{1}{2}}\right)^{-1} = (a \circ b)^{-1}$ gives the automorphic inverse property, and L is a K-loop.

Finally, unique 2-divisibility comes directly from the hypotheses, since powers are the same in L and G.

(2) For $a, b \in L$ we have

$$a^2 \circ b^2 = b^2 \circ a^2 \iff ab^2a = ba^2b \iff a^{-1}b^{-1}ab = aba^{-1}b^{-1}$$
$$\iff [a, b] = [a^{-1}, b^{-1}].$$

(3) For $a, b, c \in L$ we have

$$a^2 \circ (b^2 \circ c) = b^2 \circ (a^2 \circ c) \iff abcba = bacab$$
$$\iff [a, b]c = c[a^{-1}, b^{-1}]. \tag{ii}$$

If L is a commutative group, then the left most equation of (ii) is true for all $a, b, c \in L$. Thus with (2) we conclude $[a, b]c = c[a^{-1}, b^{-1}] = c[a, b]$.

For the converse, $[[a, b], c] = 1$ for all $a, b, c \in L$ together with (1.1) implies that $[a, b]$ is in the center of $\langle L \rangle$. Therefore

$$[a^{-1}, b^{-1}] = aba^{-1}b^{-1} = aba^{-1}b^{-1}abb^{-1}a^{-1} = ab[a, b]b^{-1}a^{-1}$$
$$= [a, b].$$

Hence (ii) together with unique 2-divisibility shows $a \circ (b \circ c) = b \circ (a \circ c)$ for all $a, b, c \in L$. Therefore, $\lambda_a\lambda_b = \lambda_b\lambda_a$ for all $a, b \in L$, and $\mathcal{M}_\ell(L)$ is a commutative group. By (1) and (6.6.3) L is A_ℓ, hence (5.9.3) implies that $L = \mathcal{N}_r(L)$ is a group. ∎

A K-loop L is called *half embedded* into the group G if there exists an injection $\eta : L \to G$ such that $\eta(L)$ satisfies the conditions in the preceding theorem, i.e.,

$$1 \in \eta(L), \quad \eta(L)^{-1} \subseteq \eta(L), \quad a\eta(L)a \subseteq \eta(L) \quad \text{for all } a \in \eta(L),$$

the map $\kappa : \eta(L) \to \eta(L); \ x \mapsto x^2$ is well-defined and bijective, and such that $\eta : L \to (\eta(L), \circ)$ is an isomorphism of the loops. The map η is called a *half embedding*.

GLAUBERMAN [40] and KREUZER [77] use the stronger requirement that L be a subset of G. Our approach seems to be more flexible, while it captures the essentials.

Notice that a K-loop which is half embedded into a group is necessarily uniquely 2-divisible by (6.14). In fact, unique 2-divisibility is also sufficient.

(6.15) Theorem. *Let L be a uniquely 2-divisible K-loop, and let $T \subseteq \mathrm{Aut}\, L$ be a transassociant of L. Then the map*

$$\eta : L \to L \times_Q T; \ a \mapsto (a^2, 1)$$

is a half embedding. In particular, L is half embedded into $\mathcal{M}_\ell$.

Proof. Recall the definition of the multiplication in $L \times_Q T$ from (2.15). As a consequence we get from (6.4.3) and (6.1.1) $(a, 1)^k = (a^k, 1)$ for all $a \in L$ and $k \in \mathbf{Z}$. In particular, $(L \times 1)^{-1} \subseteq L \times 1$. By unique 2-divisibility we can conclude that

$$\eta(L) = L \times 1 \quad \text{and} \quad \kappa : L \times 1 \to L \times 1; \ x \mapsto x^2 \quad \text{is bijective.}$$

Let $a, b \in L$. Using (2.13.1) and (6.4.1) (IV) we can compute

$$(a, 1)(b, 1)(a, 1) = (a, 1)(ba, \delta_{b,a}) = (a \cdot ba, \delta_{a,ba}\delta_{b,a})$$
$$= (a \cdot ba, 1) \in L \times 1.$$

Therefore the hypotheses of (6.14) are true for $L \times 1$, and $(L \times 1, \circ)$ becomes a K-loop, where

$$(a^2, 1) \circ (b^2, 1) = (a, 1)(b^2, 1)(a, 1) = (a \cdot b^2 a, 1) \quad \text{for all } a, b \in L.$$

It remains to show that η is a homomorphism. From (6.8.1) we obtain

$$\eta(ab) = \left((ab)^2, \mathbf{1}\right) = (a \cdot b^2 a, \mathbf{1}) = \eta(a) \circ \eta(b).$$

By (6.6.3) $T = \mathcal{D}(L)$ qualifies. Thus (2.14) shows the final statement. ■

Remarks. 1. By (6.4.1), (6.4.3) and unique 2-divisibility $\lambda(L) \subseteq \mathcal{M}_\ell$ satisfies the hypothesis of (6.14). This can be made into a direct proof for the last statement in the theorem.

2. GLAUBERMAN constructs a half embedding of L into a group isomorphic to $\mathcal{M}_\ell$ in [40; Thm. 2]. KREUZER in [77] constructs a half embedding right into $\mathcal{M}_\ell$. In fact, KREUZER's approach is a little more general, since he manages to get along with requiring the map κ to be only injective. Of course, this makes it necessary to modify the construction of the loop inside $\mathcal{M}_\ell$. With notation from (6.14), he defines $a \bullet b := (ab^2a)^{\frac{1}{2}}$, which gives a loop $(\lambda(L), \bullet)$. If L is uniquely 2-divisible, the isomorphism

$$L \times_Q \mathcal{D} \to \mathcal{M}_\ell; \ (a, \alpha) \mapsto \lambda_a \alpha$$

from (2.14) induces a K-loop $(\lambda(L), \circ)$, isomorphic to $L \times 1$. The map $(\lambda(L), \bullet) \to (\lambda(L), \circ); \ \lambda_a \mapsto \lambda_a^2$ is an isomorphism.

3. The construction in (6.14) can be done for G a Moufang loop with identical proofs, because Moufang loops are di-associative by Moufang's Theorem (see the appendix). One of the first classes of examples for K-loops was constructed this way, see [20; VII, Thm. 5.2, p. 121] and the following remark.

4. Let G be a uniquely 2-divisible group, and do the construction of (6.14) to $L := G$. This gives a big class of examples. BRUCK [20; Ex. 1, p. 124] remarks that the Burnside groups of prime exponent $p > 3$ qualify for this construction, and give (in our words) non-commutative K-loops. In particular, they are not Moufang loops. On the other hand, BRUCK even gives a Moufang loop G, which leads to a commutative group this way, [20; Ex. 3, p. 128].

5. The case $L = G$ has another interpretation: The map

$$\phi : G \to \operatorname{Aut} G; \; a \mapsto \widehat{a^{-\frac{1}{2}}}$$

is a derivation, and $a \circ b = a\phi_a(b)$. Derivations are introduced in §12.

6. (6.14.3) has been proved in [20; VII, Lemma 5.6, p. 127] for the case $L = G$, and in [40; Cor., p. 386] for GLAUBERMAN's "B-loops".

7. Frobenius Groups with Many Involutions

Before we get into Frobenius groups, we give a construction for left loops using certain sets of involutions in transitive permutation groups. This construction occurs in [55; §6] in a completely different context. It is used to coordinatize absolute spaces with K-loops. This approach has been further generalized by GABRIELI, KARZEL in [36, 37, 38], and by IM, KO in [50].

A. REFLECTION STRUCTURES

The following theorem and its proof are due to KARZEL [55].

Let G be a group acting on a set P with $J := \{\alpha \in G; \alpha^2 = 1\}$. For the fixed element $1 \in P$, denote the stabilizer by Ω.

(7.1) Theorem. For a map $\mu : P \to J; \; x \mapsto \mu_x$ with $\mu_x(1) = x$ for all $x \in P$ define

$$\lambda : \begin{cases} P \to G; \\ x \mapsto \lambda_x := \mu_x \mu_1 \end{cases}.$$

We have

(1) $L := \lambda(P)$ *is a transversal of* G/Ω. *The left loop* L *has unique inverses, and is isomorphic via* λ *to the natural left loop structure on* P. *Indeed, the* λ_x *are the left translations of* P. *Moreover,* $\mu_1(x) = x^{-1}$ *for all* $x \in P$, *where* x^{-1} *denotes the inverse of* x *in* P.

(2) L *is a loop if and only if the set* $\mu(P)$ *acts regularly on* P.

(3) *The following are equivalent*

   (I) $\mu_1 \mu(P) \mu_1 \subseteq \mu(P)$;

   (II) $\mu_1 \mu_x \mu_1 = \mu_{x^{-1}}$ *for all* $x \in P$;

  (III) L *has the automorphic inverse property;*

  (IV) L *has the left inverse property.*

(4) L *is a K-loop if and only if* $\mu_x \mu(P) \mu_x \subseteq \mu(P)$ *for all* $x \in P$.

Proof. (1) The first statement is just a special case of (2.10.3).

By (2.10.4) λ is an isomorphism. We'll make use of this isomorphism by proving the assertions for P rather than for L.

Let x' be the right inverse of $x \in P$. Then

$$xx' = \mu_x \mu_1(x') = 1, \quad \text{hence } x' = \mu_1 \mu_x(1) = \mu_1(x),$$

and

$$x'x = \mu_{\mu_1(x)} \mu_1(x) = 1, \quad \text{since } \mu_{\mu_1(x)} \text{ is an involution.}$$

Thus x' is also the left inverse of x, and $\mu_1(x) = x' = x^{-1}$.

(2) If L is a loop, then by (2.11.2) L acts regularly on P, hence so does $\mu(P)$. The converse is in (2.11.1).

(3) The last statement of (1) will be used without reference. A direct calculation shows

$$\mu_1 \lambda_x \mu_1 = \lambda_x^{-1} \quad \text{for all } x \in P.$$

(I) $\Longrightarrow$ (II): For $y \in P$ such that $\mu_1 \mu_x \mu_1 = \mu_y$ we can compute

$$x^{-1} = \mu_1(x) = \mu_1 \mu_x \mu_1(1) = \mu_y(1) = y.$$

(II) $\Longrightarrow$ (III): For $x, y \in P$, we have

$$(xy)^{-1} = \mu_1(xy) = \mu_1 \mu_x \mu_1(y) = \mu_{x^{-1}} \mu_1 \mu_1(y) = x^{-1} y^{-1}.$$

(III) $\Longrightarrow$ (IV): By (2.4), the hypothesis in (III) just means $\mu_1 \lambda_x \mu_1 = \lambda_{\mu_1(x)} = \lambda_{x^{-1}}$. Therefore $\lambda_{x^{-1}} = \lambda_x^{-1}$, and (3.1.1) shows the result.

(IV) $\Longrightarrow$ (I): For $x \in P$, (3.1.1) shows $\lambda_x = \lambda_{x^{-1}}^{-1}$. Hence we have

$$\mu_1 \mu_x \mu_1 = \mu_1 \lambda_x = \mu_1 \lambda_{x^{-1}}^{-1} = \mu_1 \mu_1 \mu_{x^{-1}} \in \mu(P).$$

(4) If P is a K-loop, then for all $x, y \in P$ we have $\lambda_x \lambda_y \lambda_x = \lambda_{x \cdot yx}$, by (6.4). Using (1) and the automorphic inverse property, we compute

$$\mu_x \mu_y \mu_x = \lambda_x \mu_1 \lambda_y \mu_1 \lambda_x \mu_1 = \lambda_x \lambda_{y^{-1}} \lambda_x \mu_1$$
$$= \lambda_{x \cdot y^{-1} x} \mu_1 = \mu_{x \cdot y^{-1} x} \mu_1 \mu_1 \in \mu(P).$$

For the converse let $x, y \in P$. We have

$$\lambda_x \lambda_y \lambda_x = \mu_x \mu_1 \mu_y \mu_1 \mu_x \mu_1 \in \mu_x \mu(P) \mu_x \mu_1 = \mu(P) \mu_1 = \lambda(P).$$

Therefore (6.4) shows that P is a Bol loop. Since (I) is valid, we conclude from (3) that P is a K-loop. ∎

Remarks. 1. KARZEL calls the triple $(P, \mu(P); 1)$ with axioms as the general hypothesis of the theorem a *reflection structure.*

2. The special case of reflection structures considered in (4) of the preceding theorem has been considered by KIKKAWA, using a different language. KINYON [72] has shown the equivalence.

B. FROBENIUS GROUPS

A transitive permutation group (G, P) with $|P| \geq 2$ is called a *Frobenius group* if G does not act regularly, and only the identity of G fixes more than one point, i.e., if Ω is the stabilizer of an arbitrary point $e \in P$, then $\Omega \neq \{1\}$ acts fixed point free on $P \setminus \{e\}$. By abuse of language we will call G a Frobenius group if such a permutation representation exists.

Let K be a group with a non-trivial group $\Phi \subseteq \operatorname{Aut} K$ acting fixed point free on K. If the map

$$1 - \omega : K \to K; g \mapsto g\omega\left(g^{-1}\right) \quad \text{is bijective for all } \omega \in \Phi \setminus \{1\},$$

then (K, Φ) is called a *Ferrero pair*. The semidirect product of K with Φ will turn out to be a Frobenius group. Notice that the action of Φ on K is fixed point free. The notion of a *fibration* is introduced in §8.

(7.2) Theorem. *Let G be a group. Consider the following conditions:*

(I) G *contains a normal subgroup K and a subgroup $\Omega \neq \{1\}$ such that $\{g\Omega g^{-1}; g \in K\} \cup \{K\}$ is a fibration of G, with $\Omega \cap g\Omega g^{-1} = \{1\}$ for all $g \in K^{\#}$;*

(II) $G = K\Omega$ *with subgroups K, Ω such that $(K, \widehat{\Omega})$ is a Ferrero pair;*

(III) $G = K\Omega$ contains subgroups K and $\Omega \neq \{1\}$ such that $\widehat{\Omega}$ acts fixed point free on K;

(IV) G has a subgroup $\Omega \neq \{1\}$ such that $\Omega \cap g\Omega g^{-1} = \{1\}$ for all $g \in G \setminus \Omega$;

(V) G is a Frobenius group.

We have

(1) (I) $\Longleftrightarrow$ (II) $\Longrightarrow$ (III) $\Longrightarrow$ (IV) $\Longleftrightarrow$ (V).

(2) If G is finite, then the above conditions are all equivalent.

Proof. (1) (I) $\Longrightarrow$ (II): Let $\omega \in \Omega^{\#}$. Since K is normal in G, the map

$$1 - \widehat{\omega} : K \to K; \quad g \mapsto g\widehat{\omega}(g^{-1}) = g\omega g^{-1}\omega^{-1}$$

is well-defined. For $g, g' \in K$ assume $(1 - \widehat{\omega})(g) = (1 - \widehat{\omega})(g')$, then

$$g\omega g^{-1} = g'\omega g'^{-1} \in g\Omega g^{-1} \cap g'\Omega g'^{-1} \neq \{1\},$$

therefore $g = g'$, and $1 - \widehat{\omega}$ is injective.

For $g \in K$, there exists $h \in K, \psi \in \Omega$ such that $g\omega = h\psi h^{-1}$, since $g\omega \notin K$. Now,

$$g\omega\psi^{-1} = (1 - \widehat{\psi})(h) \in K \implies \omega = \psi, \quad \text{thus} \quad g = (1 - \widehat{\omega})(h),$$

and $1 - \widehat{\omega}$ is surjective.

For $x \in G \setminus K$ there exist $g \in K$, $\omega \in \Omega$ with $x = g\omega g^{-1} = (1 - \widehat{\omega})(g)\omega \in K\Omega$, therefore $G = K\Omega$.

(II) $\Longrightarrow$ (I): Clearly, K is normal.

Let $g \in K^{\#}$, and assume for $h \in K, \omega \in \Omega$:

$$h = g\omega g^{-1} \in K \cap g\Omega g^{-1}.$$

Then $h\omega^{-1} = (1 - \widehat{\omega})(g) \in K$, and $\omega \in K$. Since $(1 - \widehat{\omega})(\omega) = 1$, the hypothesis implies $\omega = 1$, and $h = 1$. Thus $K \cap g\Omega g^{-1} = \{1\}$.

Now assume for $\psi, \omega \in \Omega$

$$\psi = g\omega g^{-1} \in \Omega \cap g\Omega g^{-1},$$

then $\psi w^{-1} = (1 - \hat{w})(g) \in K \cap \Omega = \{1\}$, and $\psi = w = 1$. Therefore $\Omega \cap g\Omega g^{-1} = \{1\}$.

Every element of G is of the form gw, with $g \in K$ and $w \in \Omega$. If $w = 1$, then $g \in K$. If $w \neq 1$, there exists $h \in K$ such that $g = (1 - \hat{w})(h)$. Then $gw = (1 - \hat{w})(h)w = hwh^{-1} \in h\Omega h^{-1}$. Therefore, $\{g\Omega g^{-1}; g \in K\} \cup \{K\}$ is a fibration of G.

(II) $\Longrightarrow$ (III): The requirement that Ω acts fixed point free on K is equivalent with the injectivity of $1 - \hat{w} : K \to K$.

(III) $\Longrightarrow$ (IV): Clearly, $K \cap \Omega = \{1\}$. Let $g \in G \setminus \Omega$, and assume there are $w, w' \in \Omega$ such that $w = gw'g^{-1} \in \Omega \cap g\Omega g^{-1}$. Choose $h \in K, \psi \in \Omega$ such that $g = h\psi$. Plugging in for g, we obtain

$$w = h\psi w'\psi^{-1}h^{-1}, \quad \text{hence}$$

$$w(\psi w'\psi^{-1})^{-1} = (1 - \widehat{\psi w'\psi^{-1}})(h) \in K \cap \Omega = \{1\}.$$

Thus $w = \psi w'\psi^{-1}$, and $w = hwh^{-1}$, so $h = w^{-1}hw$ is a fixed point of the action of w^{-1} on K. This implies $w = 1$, since $h \neq 1$. Therefore $\Omega \cap g\Omega g^{-1} = \{1\}$.

(IV) $\Longrightarrow$ (V): The natural action of G on the left cosets of Ω in G is transitive, and not regular, because $\Omega \neq \{1\}$, see (1.3).

For $g, x, y \in G$ assume

$$x\Omega \neq y\Omega, \quad gx\Omega = x\Omega, \quad gy\Omega = y\Omega,$$

then

$$x^{-1}gx\Omega = \Omega \quad \text{and} \quad (x^{-1}gx)(x^{-1}y)\Omega = (x^{-1}y)\Omega.$$

Thus the problem is reduced to the case $x = 1$. Now $g \in \Omega$ and $y^{-1}gy \in \Omega$ implies $g \in \Omega \cap y\Omega y^{-1} = \{1\}$. Therefore only 1 has more than one fixed point, and G is a Frobenius group.

(V) $\Longrightarrow$ (IV): Let Ω be the stabilizer of a point e, say. Since G does not act regularly, $\Omega \neq \{1\}$. For $g \in G \setminus \Omega$, we have $g(e) \neq e$, therefore $\Omega \neq g\Omega g^{-1}$. Since $g\Omega g^{-1}$ is the stabilizer of $g(e)$, by (1.3.3) we must have $\Omega \cap g\Omega g^{-1} = \{1\}$.

(2) is proved in many books on group theory, e.g., [5; (35.24), p. 191].　　　　　　　　　　　　　　　　　　　　　　■

Remarks. 1. Neither of the implications " $\Longrightarrow$ " can be reversed in general. Let F be a nearfield, Ω a subgroup in F^*, and let G be the semidirect product of F with Ω, then G satisfies (III). If F is "planar", then (I) is true, if not, and if $\Omega = F^*$, then (I) does not hold. Examples of both planar and non-planar nearfields can be found in [121; I.9]. Examples with (IV), but not (III) are constructed in §§9,11.

2. The equivalence of (II) and (III) for finite groups is easy to see. The point of FROBENIUS's theorem is the existence of the normal subgroup K, which has Ω as a complement. K is called the *Frobenius kernel* of G, and Ω the *Frobenius complement*. The group G is then the semidirect product of K and Ω.

3. When G is finite, then the Frobenius representation is unique up to equivalence, see [93; 17.5, p. 186]. This also follows easily from the uniqueness of the Frobenius kernel [42; 9.2.8, p. 179].

4. Ferrero pairs have been introduced by BETSCH and CLAY and studied by CLAY (see [26]). They arose from the study of planar nearrings.

If the Frobenius group (G, P) has no kernel, and is therefore not a semidirect product, we still have a substitute: transversals, and the quasidirect product.

(7.3) *Let* (G, P) *be a Frobenius group. For a fixed* $e \in P$ *the stabilizer* Ω *of* e *is corefree. Thus for every transversal* L *of the coset space* G/Ω, *the quasidirect product* $L \times_Q \Omega$ *is well-defined, and the permutation groups* (G, P) *and* $(L \times_Q \Omega, L)$ *are equivalent.*

Proof. Ω is corefree by (IV) of (7.2). Therefore (2.16) shows that $L \times_Q \Omega$ is well-defined.

By (2.9) the permutation groups (G, P) and (G, L) are equivalent. By (2.16) (G, L) and $(L \times_Q \Omega, L)$ are equivalent.　　　　　■

A transassociant $T \neq \{1\}$ of a left loop L will be called *fixed point free* if it acts fixed point free on $L^{\#}$. Recall that by definition, transassociants fix 1. Notice that the requirement $T \neq \{1\}$ is only necessary, when L is a group (see (2.3)).

(7.4) *Let L be a left loop, and let T be a fixed point free transassociant of L. Then $(L \times_Q T, L)$ is a Frobenius group.*

Proof. By (2.13.2) $L \times_Q T$ acts transitively on L. From (2.13.3) we know that $1 \times T$ is the stabilizer of 1. Since this group is non-trivial, the action is not regular.

Take an element $(a, \alpha) \in L \times_Q T$ which has two fixed points. Since the action is transitive, using (1.3.3) it is seen that there is no loss in generality, to assume that one of the fixed points is 1. Then $a = 1$ and $\alpha(x) = x$ for some $x \in L \setminus \{1\}$. By hypothesis, $\alpha = 1$, and the only element with more than one fixed point is $(1, 1)$, the identity. ∎

Remarks. 1. This is a straightforward generalization of the construction [81; (7.2)] for loops. It also generalizes (III) $\implies$ (IV) in (7.2).

2. If L is a loop, then clearly all elements of $L \times 1$ act fixed point free on L. This need not be the case, when L is only a left loop. Indeed, if one applies the construction of (12.3) to $G := \mathbf{Z}_3$, then $L := G^\phi$ is a left loop such that $\mathcal{D}(L) = \{\pm 1\}$ is a fixed point free transassociant. Thus $L \times_Q \mathcal{D}$ is a Frobenius group. It is easy to see that the elements of $L \times 1 \setminus \{(0, 1)\} = \{(1, 1), (2, 1)\}$ are involutions, which necessarily have fixed points. In fact, $L \times_Q \mathcal{D}$ is isomorphic to $\mathcal{S}_3$. Of course, $\mathcal{S}_3$ does have a kernel, namely $\langle (1, 2, 3) \rangle$, so in a sense we only chose the transversal in a silly way.

3. There are examples where one cannot make a better choice. Specifically, according to [27] (see also [16; p. 205]), there exists a Frobenius group G with $G = \bigcup_{g \in G} g \Omega g^{-1}$ (Ω a one point stabilizer). Therefore G contains no fixed point free elements. Here no transversal of G/Ω can be a loop.

C. INVOLUTIONS

Our aim is to establish a new class of Frobenius groups, which generalizes the notion of a sharply 2-transitive group in a reasonable way. We also present the connection with other approaches, in particular with [35]. First a simple lemma.

(7.5) Let (G, P) be a Frobenius group such that the set $J := \{g \in G; g^2 = 1\}$ acts transitively on P. Then for all $x, y \in P$ with $x \neq y$ there exists a unique element $\alpha \in J$ with $\alpha(x) = y$.

Proof. The existence of α is an assumption. If $\beta \in J$ also has the property $\beta(x) = y$, then

$$\alpha\beta(x) = \alpha(y) = x \quad \text{and} \quad \alpha\beta(y) = y, \quad \text{hence} \quad \alpha\beta = 1,$$

since G is a Frobenius group. Therefore $\alpha = \beta$ is unique.   ∎

A Frobenius group (G, P) is said to have *many involutions* if $J := \{g \in G; g^2 = 1\}$ acts transitively on P, and if the one point stabilizers contain at most one involution. Note that since all one point stabilizers are conjugate, they all contain the same number of involutions.

(7.6) Theorem. Let (G, P) be a Frobenius group with many involutions. For a fixed $e \in P$ let Ω be the stabilizer of e.

(1) There exists a unique map $\mu : P \to J$ with $\mu_x(e) = x$ for all $x \in P$ such that $\mu(P) = J \setminus \{1\}$ or $\mu(P) = J$. The two cases are mutually exclusive.

(2) $\mu(P)$ acts regularly on P.

(3) $L := \mu(P)\mu_e$ is a transversal of the coset space G/Ω, which is a K-loop.

(4) $G = L \times_Q \Omega$. What's more, the permutation groups (G, P) and $(L \times_Q \Omega, L)$ are equivalent.

(5) $\mu(P) = J \setminus \{1\}$ if and only if every involution in G has exactly one fixed point. In this case we have $C_G(\mu_e) = \Omega$, and L contains no elements of order 2.

(6) If $\mu(P) = J$, then $L = J$ is of exponent two, and every involution is fixed point free.

Proof. (1) Let $x \in P$. If $x \neq e$, there exists $\mu_x \in J$ with $\mu_x(e) = x$, which is unique by (7.5). If $e = x$, then we consider two cases: Firstly, assume that there is no involution in Ω, i.e., no involution has a fixed point, then $\mu_e = 1$, and $\mu(P) = J$. Secondly, assume that there exists a (unique!) involution $\alpha \in \Omega$, then put $\mu_e := \alpha$. Through this choice, the map μ has the desired properties. Note that there is no other way to do it — μ is unique.

(2) We have seen that Ω contains the involution μ_e if $\mu(P) \neq J$. This involution has the fixed point e. Then $\mu_x \mu_e \mu_x$ has fixed point $x \in P$. Thus for every $x, y \in P$ there exists $\alpha \in \mu(P)$ with $\alpha(x) = y$, even if $x = y$ and $1 \notin \mu(P)$.

The uniqueness of α comes from (7.5).

(3) is a direct consequence of (1), (2) and (7.1), since $\mu(P)$ is obviously invariant.

(4) From (7.2) we find $\bigcap_{g \in G} g \Omega g^{-1} = \{1\}$. Thus the result follows from (2.16) and (2.9).

(5) The first statement is clear from the proof of (1). It follows immediately from (7.2), (IV), that $C_G(\mu_e) \subseteq \Omega$. For $\omega \in \Omega$ the element $\omega \mu_e \omega^{-1}$ is an involution with fixed point e. Therefore $\omega \mu_e \omega^{-1} = \mu_e$, and $\omega \in C_G(\mu_e)$.

(6) is clear from the previously proved statements, and from the fact that in left power alternative loops, we have $|a| = |\lambda_a|$ for every element a in that loop.   ∎

In the case where L is of exponent 2 (see (6) in the preceding theorem), the Frobenius group G (with many involutions) is said to have *characteristic* 2, in symbols char $G = 2$. Otherwise, we write char $G \neq 2$. The latter is the situation where involutions have fixed points.

D. SHARPLY 2-TRANSITIVE GROUPS

Recall, a permutation group (G, P) is called *sharply 2-transitive* if for all $a_1, a_2, b_1, b_2 \in P$ with $a_1 \neq a_2, b_1 \neq b_2$ there exists exactly one $\sigma \in G$ with $\sigma(a_j) = b_j$. To avoid one trivial case, we always assume $|P| \geq 2$. Notice that G is a Frobenius group.

(7.7) Let (G, P) be a sharply 2-transitive group. For fixed $e \in P$ let Ω be the stabilizer of e. Then

(1) G is a Frobenius group with many involutions.

(2) $\widehat{\Omega}$ *acts faithfully as an automorphism group on the correspond-ing K-loop L (constructed in (7.6)). Moreover, this action is reg-ular on* $L^{\#}$.

Proof. (1) Clearly, G is a Frobenius group. For distinct $x, y \in P$ there exists exactly one $\alpha \in G$ with $\alpha(x) = y$, $\alpha(y) = x$. Clearly, $\alpha \in J$, so J is transitive (see also [121; V.2, (i), p. 229]). By [121; V.2, (i), (ii), pp. 229] either no involution has a fixed point, or all of them have exactly one fixed point. If they have fixed points, then for every $x \in P$ there is exactly one involution with fixed point x. Thus G has many involutions.

(2) We'll use notation from (7.6.1). Choose $\omega \in \Omega$, then $\omega J^{\#} \omega^{-1} \subseteq J^{\#}$ and so

$$\omega \mu_e \omega^{-1} = \mu_e, \quad \text{because} \quad \omega \mu_e \omega^{-1}(e) = e.$$

Therefore, $\omega L \omega^{-1} \subseteq L$. Now (2.8.6) shows that $\widehat{\Omega}$ acts as an au-tomorphism group on L.

Since $\omega \mu_x \omega^{-1} = \mu_{\omega(x)}$ for all $x \in P \setminus \{e\}$, and Ω is regular on $P \setminus \{e\}$, the action of $\widehat{\Omega}$ on $L^{\#}$ is regular, as well. ∎

A structure $(F, +, \cdot)$ with two binary operations "$+, \cdot$" is called a *neardomain* if $(F, +)$ is a loop with unique inverses,[1] and $(F^*, \cdot)$ is a group, such that for all $a, b, c \in F$

$$0 \cdot a = 0, \quad a \cdot (b + c) = a \cdot b + a \cdot c, \quad \text{and} \quad \exists d_{a,b} \in F^* : \delta_{a,b}(c) = d_{a,b} \cdot c.$$

A *nearfield* is a neardomain with associative addition, or equiva-lently, $d_{a,b} = 1$ for all $a, b \in F$.

We put

$$T_2(F) := \left\{ \tau_{a,b} : \begin{cases} F \to F \\ x \mapsto a + bx \end{cases} ; \ a \in F, b \in F^* \right\}.$$

(7.8) *If F is a neardomain, then $T_2(F)$ is a group acting sharply 2-transitive on F.*

---

[1] 0 denotes the identity element of any additively written loop F, and $M^* := M \setminus \{0\}$ for every subset M of F.

The *Proof* is left as an exercise, see also [121; (V.1.2), p. 217]. ∎

We shall now prove the converse, i.e., every sharply 2-transitive group arises from a neardomain in this way. We'll actually prove a bit more.

(7.9) Theorem. *Let $(L, +)$ be an (additively written) loop with unique inverses and a subgroup of the automorphism group A acting regularly on $L \setminus \{0\}$. Moreover, we assume that $\mathcal{D}(L) \subseteq A$ (so A is a transassociant). Choose $1 \in L \setminus \{0\}$ and let $\nu_a \in A$ be the unique element with $\nu_a(1) = a$. For $a, b \in L$ we put*

$$a \cdot b := \begin{cases} \nu_a(b) & \text{if } a \neq 0 \\ 0 & \text{if } a = 0 \end{cases}.$$

Then $(L, +, \cdot)$ is a neardomain, and the map

$$T_2(L) \to L \times_Q A; \quad \tau_{a,b} \mapsto (a, \nu_b), \quad (a, b \in L, b \neq 0)$$

induces an equivalence of the permutation groups $(T_2(L), L)$ and $(L \times_Q A, L)$.

Proof. Clearly, $(L^*, \cdot)$ is a group, isomorphic to A. The distributive law comes from the fact that the ν_a are automorphisms of L. Let $a, b \in L$, then $\delta_{a,b} \in A$, and there exists $c \in L$ with $\delta_{a,b} = \nu_c$. Put $d_{a,b} := c$. If $b \neq 0$, then

$$\tau_{a,b}(x) = a + bx = a + \nu_b(x) = (a, \nu_b)(x) \quad \text{for all } x \in L,$$

hence $(T_2(L), L)$ and $(L \times_Q A, L)$ are equivalent. ∎

A neardomain F clearly satisfies the hypothesis of the preceding theorem, since F^* acts regularly on $F \setminus \{0\}$.

(7.10) Theorem. **(1)** *Every sharply 2-transitive group G gives rise to a neardomain F such that $T_2(F) = G$. Moreover, char $G = 2 \iff$ char $F = 2$, i.e., $1 + 1 = 0$ in F.*

(2) *Let F be a neardomain, then the neardomain constructed as in (1) from $T_2(F)$ is isomorphic to F.*

(3) *The additive loop of a neardomain is a K-loop.*

Proof. (1) follows from (7.7) and (7.9), (2) is a consequence of [121; (V.1.5), p. 218], and (3) comes from (2) and (7.6.3). ∎

Since, by (6.4.3), K-loops are left power alternative, every element in a neardomain F generates a cyclic subgroup of $(F, +)$. By left distributivity we conclude that the map

$$\eta : \mathbf{Z} \to F; \ k \mapsto 1 \cdot k := \begin{cases} 1 + 1 + \ldots + 1 & \text{if } k \geq 0 \\ -1 - 1 - \ldots - 1 & \text{if } k < 0 \end{cases} \quad (|k| \text{ times})$$

is a ring homomorphism. Let $n\mathbf{Z}$, $n \geq 0$, be the kernel. Since F has no zero-divisors, n is a prime or $n = 0$. Thus we can call n the *characteristic* of F, in symbols $\operatorname{char} F := n$ (cf. [121; p. 224]). In case $\operatorname{char} F = 0$ the map η can be extended to $\mathbf{Q}$, the field of fractions of $\mathbf{Z}$, in the well-known way. Thus in any case F contains a *prime field* isomorphic to $\mathbf{Z}_p$ or $\mathbf{Q}$ if $\operatorname{char} F = p$ or $\operatorname{char} F = 0$, respectively.

If P is the prime field of F, then for all $q, q' \in P$, $a \in F$ we have $a(q + q') = aq + aq'$. This implies (see also [59])

(7.11) *Let F be a neardomain with $\operatorname{char} F = p$, then the order of every non-zero element in $(F, +)$ is p or infinity if $p = 0$. In particular, if $p \neq 2$, then $(F, +)$ is uniquely 2-divisible.* ∎

Remarks. 1. It seems to be still open, whether neardomains with non-associative addition exist. The theorem shows a potential path for a construction starting from K-loops.

2. Zassenhaus [124] (see also Karzel [53]) showed that all finite neardomains are nearfields. Tits [113, 114] showed that locally compact, connected neardomains are nearfields, and Kerby and Wefelscheid [59] showed that neardomains of characteristic 3 are nearfields. For more results of a similar kind cf. [121; V.1.B].

3. For the study of sharply 3-transitive groups special neardomains, called *KT-fields*, have been introduced by Kerby and Wefelscheid [60]. Kerby [58] showed that every KT-field F with $\operatorname{char} F \equiv 1 \mod 3$ is a nearfield.

E. Characteristic 2

For Frobenius groups with many involutions of characteristic 2 we obtain a satisfactory converse to (7.6). At the same time, we get a considerable strengthening of theorem (3.4) for loops of exponent 2.

(7.12) Theorem. *Let L be a loop of exponent 2, and let T be a fixed point free transassociant. Then T contains no involutions, $G := L \times_Q T$ is a Frobenius group with many involutions and char $G = 2$. Moreover, L is a K-loop, and $T \subseteq \operatorname{Aut} L$.*

Proof. G is a Frobenius group by (7.4), and the set $L \times 1$ acts transitively. By (3.4) L satisfies the left inverse property. Thus for all $a \in L$

$$\delta_{a,a} = 1 \quad \text{and so} \quad (a, 1)(a, 1) = (a^2, \delta_{a,a}) = (1, 1).$$

Therefore, $L \times 1$ consists of involutions and the identity.

We identify T and $1 \times T$ according to (2.13). Assume, there exists an involution $\alpha \in T$. Then for $x \in L \setminus \{1\}$ we have $\alpha(x) = y \neq x$. Take $a \in L$ with $ax = y$. Then $g := (a, 1) \in G$ is an involution, and $g(x) = y$. By (7.5) we arrive at the contradiction $g = \alpha$. Therefore T contains no involutions. This also implies that G is a Frobenius group with many involutions, and char $G = 2$.

By (7.6.3) $L \times 1$ is a K-loop, which is isomorphic to L (2.13.4). For $a \in L, \gamma \in T$ we have

$$(1, \gamma)(a, 1)(1, \gamma)^{-1} = \big(\gamma(a), \chi(a, \gamma)\big) \in L \times 1,$$

since the conjugate of an involution is an involution. Therefore, $\chi(a, \gamma) = 1$ for all $a \in L$, and γ is an automorphism by (2.12.3). Hence $T \subseteq \operatorname{Aut} L$. ∎

It would be desirable to relax the definition of "Frobenius group with many involutions" to the sole requirement that the set J acts transitively. This is possible in some special cases.

(7.13) *Let (G, P) be a Frobenius group such that the set $J := \{g \in G;\ g^2 = 1\}$ acts transitively on P. Take Ω to be the stabilizer of a fixed $e \in P$.*

(1) *If there exists a subset F of J, $1 \in F$, which acts fixed point free and transitive, then G is a Frobenius group with many involutions and char $G = 2$. Furthermore, $F = J$, and F is a transversal of the coset space G/Ω, which is a K-loop.*

(2) *If $G = K\Omega$ with a Ferrero pair (K, Ω), then G is a Frobenius group with many involutions. Furthermore, K is an abelian group.*

Proof. (1) From (7.5) and the assumption we conclude that F acts regular. Therefore F is a loop transversal of G/Ω by (2.11.1). Moreover, Ω is a fixed point free transassociant of F, since Ω is corefree, and G is Frobenius. By (3.4) F satisfies the left inverse property. Therefore, F is of exponent 2. Now (7.12) applies, and gives all the assertions, because $G = L \times_Q \Omega$ by (2.16).

(2) For a first case, assume Ω contains no involutions. Then $\bigcup_{g \in G} g \Omega g^{-1} \cap J = \{1\}$. In view of (7.2)(I) we have $J \subseteq K$. Therefore (1) applies and gives the first statement.

To see that K is an abelian group observe that J acts transitively, and K acts regularly on K. This forces $J = K$, and K is of exponent 2, hence abelian by (1.2).

If $\varepsilon \in \Omega \cap J^{\#}$, then we have for all $g \in K$

$$\hat{\varepsilon}([g, \varepsilon]) = \varepsilon g \varepsilon g^{-1} \varepsilon \varepsilon = \varepsilon g \varepsilon g^{-1} = [g, \varepsilon]^{-1}.$$

By hypothesis every element in K is of the form $[g, \varepsilon]$ for some $g \in K$. Hence $\hat{\varepsilon}(g) = g^{-1}$ for all $g \in K$. Since $\hat{\varepsilon}$ is an automorphism of K, K must be abelian. ε was arbitrary, and we saw it must satisfy $\hat{\varepsilon}(g) = g^{-1}$ for all $g \in K$. Since the map $\varepsilon \mapsto \hat{\varepsilon}$ is injective, there is exactly one involution in Ω. Hence G has many involutions. ∎

The theorem covers, in particular, all finite Frobenius groups (see [93; 18.1, p. 193]). No decisive results have been achieved in cases not mentioned in the theorem. So we do not know if there exist Frobenius groups with "too many" involutions, i.e., such that the set J acts transitively, and a one point stabilizer contains more than one involution.

F. CHARACTERISTIC NOT 2 AND SPECIFIC GROUPS

We now turn to the case of characteristic $\neq 2$. Here we get only a partial converse of (7.6).

(7.14) Theorem. *Let L be a uniquely 2-divisible K-loop, and let Ω be a fixed point free subgroup of $\operatorname{Aut} L$ which contains $\mathcal{D}(L)$ and an involution ι. Then $L \times_Q \Omega$ is a Frobenius group with many* *involutions of characteristic $\neq 2$. Moreover, for all $a, x \in L$ we have $\iota(x) = x^{-1}$, (a, ι) is an involution, and $\lambda_a(x) = (a, \iota)(1, \iota)(x)$.*

Proof. By (2.12.3) Ω is a transassociant. Therefore, (7.4) shows that $L \times_Q \Omega$ is a Frobenius group. From (6.11) it follows that ι is the map $x \mapsto x^{-1}$. Thus Ω contains exactly one involution.

A simple calculation shows that (a, ι) is an involution in $L \times_Q \Omega$. Therefore, the involutions act transitively. More specific: Let $x, y \in L$ and choose $a \in L$ such that $ax^{-1} = y$. Then $(a, \iota)(x) = y$.

Finally, $(a, \iota)(1, \iota)(x) = a(x^{-1})^{-1} = ax = \lambda_a(x)$.   ■

The following lemma shows that the presence of an involution in Ω is not a strong assumption. Indeed, it can always be achieved.

(7.15) *Let L be a K-loop, and let Ω be a subgroup of Aut L with $\iota \notin \Omega$. Then*

$$\langle \Omega \cup \{\iota\} \rangle = \Omega \times \langle \iota \rangle.$$

If L is uniquely 2-divisible, and Ω acts fixed point free, then so does $\Omega \times \langle \iota \rangle$.

Proof. Since ι centralizes every automorphism, the first statement is clear.

Now, assume that L is uniquely 2-divisible, and Ω acts fixed point free. Every element of $\Omega \times \langle \iota \rangle$ is of the form ω or $\omega\iota$ for some $\omega \in \Omega$. If this element is not the identity, and has a fixed point $x \in L \setminus \{1\}$, then it must be of the form $\omega\iota$, because of the hypothesis. Now $x = \omega\iota(x) = \omega(x^{-1})$ implies $\omega^2(x) = x$, because $\omega(x^{-1}) = \omega(x)^{-1}$. Therefore $\omega^2 = 1$. By (6.11) we necessarily have $\omega = 1$, since $\omega = \iota$ contradicts our assumption. However, $x = \iota(x) = x^{-1}$ is a contradiction, too, because a uniquely 2-divisible loop cannot have elements of order 2.   ■

Remarks. 1. If L is a uniquely 2-divisible K-loop, then Ω (as in (7.14)) cannot contain more than one involution by (6.11). If Ω happens to have no involution, then $L \times_Q \Omega$ can be embedded into the Frobenius group $L \times_Q (\Omega \times \langle \iota \rangle)$.

2. In (11.3.2) we give examples of K-loops with $\iota \notin \mathcal{D}(L)$. All of these are uniquely 2-divisible. Therefore, the hypothesis about ι in (7.14) is not redundant.

3. The converse in theorem (7.14) is only partial, because we need the assumption that L be uniquely 2-divisible. (7.6.5) together with (6.8.2) only implies that the map $x \mapsto x^2$ is injective. Surjectivity is missing. There are in fact counterexamples arising from (9.3.5), since there exist pythagorean fields, which are not euclidean (see also (1.7)).

(7.16) *Let G be a group with set of involutions $J^\#$, and let Ω be a subgroup of G. The following are equivalent*

   (I)   *$(G, G/\Omega)$ is a Frobenius group with many involutions, and char $G \neq 2$;*

   (II) *$\Omega = C_G(\mu)$ for some $\mu \in J^\#$, and for all $\alpha, \beta \in J^\#, \alpha \neq \beta$, there exists $\gamma \in J^\#$ with $\gamma \alpha \gamma = \beta$, and $C_G(\alpha) \cap C_G(\beta) = \{1\}$.*

If (II) holds, then γ is uniquely determined given α and β.

Proof. (I) $\implies$ (II): From (7.6.5) we have that $\Omega = C_G(\mu)$ for some $\mu \in J^\#$.

By assumption, α and β each have exactly one fixed point, denote them by x, y, respectively, i.e., $\alpha x = x$, $\beta y = y$. Since the involutions act transitively, there exists $\gamma \in J^\#$ with $\gamma x = y$. Now, $\gamma \alpha \gamma$ is an involution with fixed point y. Since there is only one such involution, we must have $\gamma \alpha \gamma = \beta$.

Finally, there exist $g, h \in G$ with $g \mu g^{-1} = \alpha$, $h \mu h^{-1} = \beta$, therefore, using (IV) from (7.2), we find

$$C_G(\alpha) \cap C_G(\beta) = C_G(g \mu g^{-1}) \cap C_G(h \mu h^{-1})$$
$$= g \Omega g^{-1} \cap h \Omega h^{-1} = \{1\}.$$

Indeed, the case $h^{-1} g \in \Omega$ leads to $\alpha = \beta$, which is excluded by hypothesis.

(II) $\implies$ (I): All the conjugates of Ω are centralizers of involutions, thus (IV) of (7.2) is satisfied, and $(G, G/\Omega)$ is a Frobenius group.

To see that $J^\#$ acts transitively, take $X, Y \in G/\Omega$. The stabilizer of $\Omega \in G/\Omega$ is Ω itself, and contains the involution μ. Hence every one point stabilizer contains an involution, because they are conjugate to Ω. In particular, there exists an involution α with fixed point X. Also, there exists an element $g \in G$ with $gX = Y$. Now gag^{-1} is an involution, therefore there exists $\gamma \in J^{\#}$ with $\gamma(gag^{-1})\gamma = \alpha$. Hence $\gamma g\alpha = \alpha\gamma g$, and $\gamma g \in C_G(\alpha)$, the stabilizer of X. This implies $\gamma \in gC_G(\alpha)$, and so $\gamma X = Y$. Thus $J^{\#}$ acts transitively.

The assumption clearly implies that there is at most one involution in $C_G(\alpha)$. Thus the Frobenius group G has many involutions. Clearly, char $G \neq 2$.

For the last statement, let $\gamma\alpha\gamma = \beta$. If $\alpha = \beta$, then $\gamma\alpha\gamma\alpha = 1$. Hence γ and α commute, i.e., $\gamma \in C_G(\alpha)$, therefore $\gamma = \alpha$. This shows uniqueness in the case of equality.

If $\alpha \neq \beta$, let $\gamma' \in J^{\#}$ also have the property $\gamma'\alpha\gamma' = \beta$. Then

$$\gamma\alpha\gamma = \gamma'\alpha\gamma' \implies \gamma'\gamma\alpha = \alpha\gamma'\gamma \quad \text{and, similarly} \quad \gamma'\gamma\beta = \beta\gamma'\gamma.$$

Therefore, $\gamma'\gamma \in C_G(\alpha) \cap C_G(\beta) = \{1\}$, and $\gamma = \gamma'$ is unique. ∎

The proof of the preceding theorem has been inspired by [35], in particular § 3.2, p. 20f. Besides its intrinsic interest, this theorem will serve to establish the connection with GABRIEL's thesis.

In [35], GABRIEL has introduced *specific groups* in order to axiomatize the subgroup of a sharply 2-transitive group generated by the set of involutions. We give a generalized notion, which includes the whole sharply 2-transitive group, and more. A group G is called *specific* if the set $J^{\#}$ of involutions in G has at least 2 elements, i.e., $|J^{\#}| \geq 2$, and there exists $p \in \mathbf{N} \cup \{\infty\}$ such that for all $\alpha, \beta \in J^{\#}$ with $\alpha \neq \beta$, there is a $\gamma \in J^{\#}$ with

$$|\alpha\beta| = p, \quad \gamma\alpha\gamma = \beta \quad \text{and} \quad C_G(\alpha) \cap C_G(\beta) = \{1\}.$$

Notice that G cannot be abelian, since two distinct involutions do not commute. As mentioned earlier, our definition slightly deviates from GABRIEL's. In fact, in [35; p. 6] it is required that G is non-abelian, and that $J^{\#}$ generates G. It is then derived that there are at least 2 involutions. This is used to show that p is either an odd prime, or ∞, [35; 2.4, p. 6]. Therefore, this statement holds in our more general situation. Call

$$\text{char } G := \begin{cases} 0 & \text{if } p = \infty \\ p & \text{otherwise} \end{cases}$$

the *characteristic* of G. For completeness, we give a proof for

(7.17) *The characteristic of a specific group is 0 or an odd prime.*

Proof. Let p be finite, and assume $p = mn$, with $n < p$. Then $(\alpha\beta)^p = \left(((\alpha\beta)^m\alpha)\alpha\right)^n$, and $(\alpha\beta)^m\alpha$ is the conjugate of an involution, hence it is an involution. Since $n < p$ we must have $(\alpha\beta)^m\alpha = \alpha$. Therefore $m = p$ and $n = 1$. Thus p is a prime.

Moreover, $p \neq 2$, since otherwise, $\alpha\beta = \beta\alpha$, contradicting our assumption.   ∎

We now hook up the notion of a specific group to the main subject of this section.

(7.18) Theorem. *Let G be a specific group of characteristic p. If μ is an involution, then $\left(G, G/\mathcal{C}_G(\mu)\right)$ is a Frobenius group with many involutions, and* char $G \neq 2$. *Moreover, the K-loop L coming with G is of exponent p, or in case $p = 0$, every element in $L^{\#}$ has infinite order.*

Proof. Let $J^{\#}$ be the set of involutions. By (7.16) G is a Frobenius group with many involutions. For $a \in L$ we have $\lambda_a = \alpha\beta$ with $\alpha, \beta \in J^{\#}$ from (7.6.3). Therefore $|\lambda_a| = p$, or $|\lambda_a| = \infty$, according to the cases in the hypothesis. But $|\lambda_a|$ is the order of a in L, because L is left power alternative, see (6.4.3) and (6.1).   ∎

This theorem has the following converse

(7.19) Theorem. *Let L be a K-loop, and let Ω be a fixed point free subgroup of* Aut L *which contains $\mathcal{D}(L)$ and ι. Put $G := L \times_Q \Omega$.*

(1) *If L is of exponent p, where p is an odd prime, then G is a specific group with* char $G = p$.

(2) *If every element in $L^{\#}$ has infinite order, then G is a specific group with* char $G = 0$.

(3) *In both cases (1) and (2), $\Omega = \mathcal{C}_G(\iota)$.*

Proof. The assumptions about the orders of elements in L imply that L is uniquely 2-divisible. Thus by (7.14) $(G, G/\Omega)$ is a Frobenius group with many involutions. From (7.6.5) we infer that $\Omega = \mathcal{C}_G(\iota)$. Now (7.16) gives all the conditions for specific groups except for the statement about $|\alpha\beta|$. So let α, β be involutions in

G. There exists an involution $\gamma \in G$ with $\beta = \gamma\iota\gamma$. Now

$$\alpha\beta = \gamma(\gamma\alpha\gamma)\iota\gamma \implies |\alpha\beta| = |(\gamma\alpha\gamma)\iota| = \begin{cases} p & \text{in case (1)} \\ \infty & \text{in case (2)} \end{cases}$$

by (7.14) and the hypothesis. This shows the remaining condition. ∎

As a corollary we get

(7.20) *Let F be a neardomain of characteristic $p \neq 2$, and let K be the subgroup of $T_2(F)$ generated by the involutions in $T_2(F)$. For G a subgroup of $T_2(F)$ containing K, i.e., $K \subseteq G \subseteq T_2(F)$, we have*

(1) *G is a specific group of characteristic p.*

(2) *$G = F \times_Q \Omega$, where Ω is the stabilizer of 1 inside G. The map $\tau_{0,.} : F^* \to T_2(F)$; $b \mapsto \tau_{0,b}$ is a monomorphism with Ω in its image. Moreover, $\Omega' := \tau_{0,.}^{-1}(\Omega)$ is a subgroup of F^* with $-1 \in \Omega'$ and $d_{a,b} \in \Omega'$ for all $a, b \in F$.*

Proof. (2) All the maps $\tau_{a,1}$ are in G, because they are products of two involutions by (7.9), (7.11), and (7.14). Therefore (G, F) is a Frobenius group with many involutions, since $T_2(F)$ is (see (7.7)). Now (7.6.4) shows the first result.

By definition, $\tau_{0,.}$ is a monomorphism, and $\tau_{0,.}(F^*)$ is the stabilizer of 1 inside $T_2(F)$. Therefore the statements about Ω follow easily.

(1) By (7.11) $(F, +)$ is uniquely 2-divisible and of exponent p or every element has infinite order, according to $p \neq 0$ or $p = 0$, respectively. Thus (2) and (7.19) give the result. ∎

Remarks. 1. In (7.19), the involutions generate G if and only if $\Omega = \langle \mathcal{D} \cup \{\iota\} \rangle$, the minimal possible choice for Ω satisfying the hypothesis if Ω acts fixed point free.

2. The fact that specific groups are Frobenius groups has been observed by GABRIEL in [35; 4.2, p. 27]. He has used the phrase "generalized Frobenius group" to denote what we call a Frobenius group.

3. With our definition every intermediate group between K and $T_2(F)$ is a specific group. In GABRIEL's setting only K qualifies, see [35; 2.4, p. 20] and remark 1. This shows the extent to which GABRIEL's notion has been generalized.

4. The proof of (7.19) shows that the definition of "specific group" can be relaxed to $|\alpha\beta| = p$ for fixed β.

5. (7.20.2) has a converse: If Ω is a subgroup of the multiplicative group of a neardomain F, with -1, $d_{a,b} \in \Omega$ for all $a, b \in F$, then $G := F \times_Q \tau_0, . (\Omega)$ is a specific group sitting between K and $T_2(F)$, and char $G = $ char F.

6. Examples of specific groups of characteristic 0 abound, see (11.3.2) and (9.3.3). This seems different in characteristic $p > 2$. The only examples known to the author arise from infinite Burnside groups as described in [35; 5.18, p. 50].

7. There are also plenty of Frobenius groups with many involutions in characteristic 2, see (11.3.1).

8. Loops with Fibrations

Let L be a loop. A set $\mathcal{F}$ of non-trivial subloops of L is called a *fibration* if

$$|\mathcal{F}| \geq 2, \quad L = \bigcup_{F \in \mathcal{F}} F, \quad \text{and} \quad \forall F_1, F_2 \in \mathcal{F} : F_1 \cap F_2 = \{1\}.$$

The pair $(L, \mathcal{F})$ will be called a *loop with fibration*, or a *fibered loop*. Let T be a subset of $\mathcal{S}_L$, then $\mathcal{F}$ is called *T-invariant* if for all $\alpha \in T$, $F \in \mathcal{F}$, we have $\alpha(F) \in \mathcal{F}$.

Obviously, $\mathcal{D}$-invariant fibrations can be characterized by the precession maps:

(8.1) Let $(L, \mathcal{F})$ be a fibered loop. $\mathcal{F}$ is $\mathcal{D}$-invariant if and only if for all $a, b \in L$, $F \in \mathcal{F} : \delta_{a,b}(F) \in \mathcal{F}$. ∎

With a fibered loop $(L, \mathcal{F})$ one can associate an incidence structure by putting

$$\mathcal{L}(\mathcal{F}) := \{aF; a \in L, F \in \mathcal{F}\}.$$

A loop L together with a set of lines $\mathcal{L}$ will be called an *incidence loop* if $(L, \mathcal{L})$ is an incidence space, and if $\lambda(L)$ is contained in the automorphism group of $(L, \mathcal{L})$, i.e., $\mathcal{M}_\ell \subseteq \text{Aut}(L, \mathcal{L})$.

ZIZIOLI has studied fibered loops and their geometries. She obtained [125; (8),(9)]:

(8.2) Let L be a loop with fibration $\mathcal{F}$. Then $(L, \mathcal{L}(\mathcal{F}))$ is an incidence loop if and only if $\mathcal{F}$ is $\mathcal{D}$-invariant. ∎

Remark. ZIZIOLI introduced the notion of a *fibered incidence loop*, i.e., an incidence loop such that the lines stem from a fibration. This notion is equivalent with our $\mathcal{D}$-invariant fibration.

If $(L, \mathcal{F})$ is a fibered loop, define a relation $\|$ on $\mathcal{L} := \mathcal{L}(\mathcal{F})$ by

$$A \parallel B, \ A, B \in \mathcal{L} \iff \exists a, b \in L, D \in \mathcal{F} \text{ with } A = aD, B = bD.$$

(8.3) Theorem. Let L be a loop with a $\mathcal{D}$-invariant fibration $\mathcal{F}$.

(1) For $F \in \mathcal{F}$ the following are equivalent

    (I) $\delta_{a,x}(F) = F$ for all $a \in L$, $x \in F$;

(II) $aF \cap bF \neq \varnothing \implies aF = bF$ for all $a, b \in L$;

(III) $a \in bF \implies aF = bF$ for all $a, b \in L$.

(2) If the conditions in (1) hold for all $F \in \mathcal{F}$, then for $E \in \mathcal{F}$, $a, b \in L$ with $aE \subseteq bF$ we have $E = F$ and $aF = bF$. This implies that every $A \in \mathcal{L}(\mathcal{F})$ has a unique $F \in \mathcal{F}$ with $A = aF$, for appropriate $a \in L$.

(3) The conditions in (1) are satisfied for all $F \in \mathcal{F}$ if and only if $(L, \mathcal{L}(\mathcal{F}), \|)$ is an incidence space with parallelism.

Proof. (1) (I) $\implies$ (II): Let $c \in aF \cap bF$. Then there exists $x \in F$ such that $ax = c$. Now $cF = ax \cdot F = a \cdot x\delta_{a,x}^{-1}(F) = a \cdot xF = aF$, and similarly $cF = bF$.

(II) $\implies$ (III): Since $a \in bF$, we have $aF \cap bF \neq \varnothing$, hence $aF = bF$.

(III) $\implies$ (I): For $a \in L$, $x \in F$ we have $ax \in aF$. Thus

$$ax \cdot F = aF = a \cdot xF \quad \text{and} \quad \delta_{a,x}(F) = F.$$

(2) Since $a \in aE$, we have $aF = bF$. Thus $aE \subseteq aF$ and canceling a gives $E \subseteq F$. But $\mathcal{F}$ is a fibration, therefore $E = F$.

(3) see [125; (10)].   ∎

If the left multiplications in P are automorphisms of $(P, \mathcal{L}, \|)$, ZIZIOLI gets a rather strong result [125; (11)]. Part (2) is implicit in the proof of [125; (13)], but will be proved here for completeness.

(8.4) Theorem. Let $(P, \mathcal{L}(\mathcal{F}), \|)$ be an incidence space with parallelism, then

$$\lambda(P) \subseteq \text{Aut}(P, \mathcal{L}(\mathcal{F}), \|) \iff \forall a, b \in P, F \in \mathcal{F} : \delta_{a,b}(F) = F.$$

In this case we have

(1) $\mathcal{M}_\ell$ consists of dilatations.

(2) If P is a proper loop, the group of dilatations, as well as $\mathcal{M}_\ell$ act on P as Frobenius groups.

(3) If P is finite, then P is a group.

Proof of (2). Clearly, $\mathcal{M}_\ell$ acts transitively. If P is a proper loop, then $\lambda(P) \neq \mathcal{M}_\ell$, and $\mathcal{M}_\ell$ does not act regularly, neither does the group of dilatations. Now (1.4) together with (1) shows the result.

(3) This is [125; (13)]. It can be proved using FROBENIUS's theorem (7.2.2). ∎

We'll now describe some examples. Specific instances with exponent 2 of the next theorem are in (11.3.1).

(8.5) *Let L be a left power alternative loop of prime exponent, which is not a cyclic group. Then $\mathcal{F} := \{\langle a \rangle ; a \in L^{\#}\}$ is an Aut L-invariant fibration of L. In particular, if L is an A_ℓ-loop, then $\mathcal{F}$ is $\mathcal{D}$-invariant.*

Proof. $|\mathcal{F}| \geq 2$, since L is not a cyclic group. Clearly, $L = \bigcup_{F \in \mathcal{F}} F$. Assume there exists $a \in L$, $a \neq 1$, and $F_1, F_2 \in \mathcal{F}$ with $a \in F_1 \cap F_2$. Now, $\langle a \rangle$, F_1 and F_2 are cyclic groups of the same finite (prime) order. Hence $F_1 = \langle a \rangle = F_2$.

For every $\alpha \in$ Aut L, $a \in L$, we have $\alpha(\langle a \rangle) = \langle \alpha(a) \rangle$, therefore $\mathcal{F}$ is Aut L-invariant. The last statement is now obvious. ∎

More involved is the following construction, which is due to KOLB and KREUZER [75]. Let L be a loop. Define a relation

$$a \sim b \quad \text{if} \quad \delta_{a,b} = 1, \quad \text{and put} \quad [a] := \{x \in L; a \sim x\}.$$

We have

(8.6) Theorem. *Let L be a Kikkawa loop. Then the relation $\sim$ as above is symmetric.*

(1) *$\sim$ is reflexive if and only if L is left alternative.*

(2) *$\sim$ is transitive if and only if for all $a, x, y \in L^{\#}$*

$$\delta_{a,x} = \delta_{a,y} = 1 \implies \delta_{x,y} = 1.$$

In this case L is left power alternative, hence $\sim$ is an equivalence relation. The corresponding classes $[a]$ are commutative subgroups of L. Moreover, if L is not a group, then $\mathcal{F} := \{[a]; a \in L\}$ is a Aut L-invariant (and therefore $\mathcal{D}$-invariant) fibration of L.

Proof. Symmetry comes from (3.7.3), (1) is clear.

(2) The first statement is obvious, given the symmetry of $\sim$.

Take $a \in L^{\#}$. By the left inverse property we have $\delta_{a^{-1},a} = 1$, thus $\delta_{a,a} = 1$, and L is left alternative. Therefore $\sim$ is an equivalence relation with $a^{-1} \in [a]$. For $b, c \in [a]$, $b \neq 1$, (3.7.2) shows

$$1 = \delta_{b,c} = \delta_{b^{-1},bc}, \quad \text{hence} \quad bc \in [b^{-1}] = [a].$$

By construction the multiplication restricted to $[a]$ is associative, hence $[a]$ is a subgroup of L. The automorphic inverse property makes it commutative.

If L is not associative, then there exist $a, b \in L$ with $\delta_{a,b} \neq 1$ by (2.3). Hence $[a] \neq [b]$ and $\mathcal{F}$ is a fibration. Finally, (2.4.2) shows that $\sim$ is $\text{Aut}\, L$-invariant, i.e., for all $\alpha \in \text{Aut}\, L$, $a, b \in L$, $a \neq 1$, we have

$$a \sim b \implies \alpha(a) \sim \alpha(b). \quad \text{Hence} \quad \alpha([a]) = [\alpha(a)].$$

This shows that $\mathcal{F}$ is $\text{Aut}\, L$-invariant.

Since $a^k \in [a]$ for all $a \in L^{\#}$, $k \in \mathbf{Z}$ we have $\delta_{a,a^k} = 1$. By (6.1.1) L is left power alternative. ∎

Remarks. 1. If $\sim$ is an equivalence relation, and $\mathcal{N}_\ell(L) \neq \{1\}$, then L is a group. This follows directly from the displayed proposition in the theorem and (5.1.1).

2. Of course, K-loops satisfy all the hypotheses of the theorem, except possibly the displayed condition which makes sure that $\sim$ is transitive. In view of (6.9.2) our loops are pretty close to K-loops. It seems open whether the conditions might imply that L is a K-loop already. The counterexamples to theorem (6.9) given in (12.3.5) are not suitable, because it is easy to see that for every η-derivation $\ker \eta$ is contained in $\mathcal{N}_\ell$, hence $\mathcal{N}_\ell \neq \{1\}$. See also [62; (2.12), p. 36] or [63; (1.3)].

(8.7) *Let L be a Kikkawa loop with fixed point free $\mathcal{D}$. Assume that $[a]$ is a subgroup of L for all $a \in L^{\#}$. Then $\sim$ is an equivalence relation.*

Proof. Take $x, y \in L$ with $\delta_{a,x} = \delta_{a,y} = 1$. Then $x, y, xy \in [a]$, so $\delta_{a,xy} = 1$ as well. By (3.7.3) we have

$$xa = \delta_{a,x}(xa) = ax, \quad ya = ay \quad \text{and} \quad xy \cdot a = a \cdot xy.$$

Using $\delta_{x,a} = 1$, we can compute

$$xy \cdot a = a \cdot xy = ax \cdot y = xa \cdot y = x \cdot ay = x \cdot ya = xy \cdot \delta_{x,y}(a).$$

This implies $\delta_{x,y}(a) = a$. The assumptions enforce $\delta_{x,y} = 1$, and (8.6.2) shows the result. ∎

9. K-Loops from Classical Groups over Ordered Fields

In this section we present a class of examples which are derived from the classical groups over an ordered field R and the extension $R(i)$. It gives a unified way to construct many examples in the literature. A major part of this has been published in [64].

A. THE CONSTRUCTION

Let $n \geq 2$ be a fixed integer, assume R is an n-real field, and set $K := R(i)$. Let G be a subgroup of $\mathrm{GL}(n, K)$. We'll use notation introduced in §1. Explicitly, we recall

$$L_G := G \cap \mathcal{H}(n, K), \quad \Omega_G := G \cap \mathrm{U}(n, K).$$

If $G = L_G \Omega_G$, then from (1.15) and (1.11) we know that L_G is a transversal, hence a left loop by (2.7). Recall that the multiplication "$\circ$" is given as follows

$$\forall A, B \in L_G : \exists_1 A \circ B \in L_G, \exists_1 d_{A,B} \in \Omega_G : AB = (A \circ B)d_{A,B}.$$

This is just the polar decomposition of AB. Also recall that

$$\kappa : \mathcal{H}(n, K) \to \mathcal{H}(n, K); \quad X \mapsto X^2$$

was the squaring map. By (1.14) κ is injective. So it makes sense to write $\kappa^{-1}(X)$ when X is a square in $\mathcal{H}(n, K)$.

(9.1) Theorem. *Let R be n-real, and let G be a subgroup of $\mathrm{GL}(n, K)$ with $G = L_G \Omega_G$, then*

(1) *$(L_G, \circ)$ is a K-loop such that inverses of elements in L_G formed in $(L_G, \circ)$ and in G coincide.*

(2) *For all $\omega \in \Omega_G$ we have $\omega L_G \omega^{-1} = L_G$, and $\delta_{A,B} = \hat{d}_{A,B}$ for all $A, B \in L_G$.*

(3) *The natural permutation representation of G on G/Ω_G induces a monomorphism $\Omega_G / \mathcal{C}_{\Omega_G}(L_G) \to \mathrm{Aut}\, L_G$. The image contains $\mathcal{D}(L_G)$.*

(4) $A \circ B = \kappa^{-1}(AB^2A)$, therefore $\det d_{A,B} = 1$ for all $A, B \in L_G$. Moreover, integer powers of $A \in L_G$ formed in $(L_G, \circ)$ and in G coincide.

(5) For all $\alpha \in \operatorname{Aut} G$ with $\alpha(L_G) \subseteq L_G$ we have $\alpha(d_{A,B}) = d_{\alpha(A),\alpha(B)}$.

Proof. (1) We have already seen that L_G is a transversal. From (1.11) and the fact that $A^* = A$ we find $AL_GA = L_G$ for all $A \in L_G$. Thus (3.12) shows that L_G is a Bol loop.

Now again (1.11) implies $L_G^{-1} \subseteq L_G$. From (3.3.1) we get the statement about the inverses. Moreover, we can use (3.3.2) to show the automorphic inverse property. Indeed, for $A, B \in L_G$ we compute

$$ABd_{A,B}^{-1} = (ABd_{A,B}^{-1})^* = (d_{A,B}^{-1})^* B^* A^* = d_{A,B}BA,$$

which is exactly what we need.

(2) The first assertion is direct from (1.11), since $\omega^* = \omega^{-1}$, the second comes from (2.8.3) and (2.8.6).

(3) is direct from (2.8.6).

(4) Using the displayed identity from the proof of (1) and the definition, we find

$$(A \circ B)^2 = ABd_{A,B}^{-1}d_{A,B}BA = AB^2A.$$

Therefore, (1.14) gives the first statement. Now, $\det A \circ B = \det AB$, since both are positive. Thus $\det d_{A,B} = 1$.

The last statement is an easy consequence of the first, and of (1).

(5) We have $d_{A,B} = (\kappa^{-1}(AB^2A))^{-1}AB$, by (4) and the definition. This implies the assertion. ∎

Remarks. 1. When R is the field of real numbers, the above theorem can be generalized to Hilbert spaces H over R or $R(i)$. Here, G is the unit group of the Banach algebra of bounded operators $H \to H$. In this case the polar decomposition is proved in [98; Thm. VIII.6, p. 73] with L the set of positive (self-adjoint) operators in G. The details will be left to the reader.

2. Part (5) is a generalization of (2.7.6), because the hypothesis "$\alpha(\Omega_G) \subseteq \Omega_G$" can be dispensed with.

We shall now go through various subgroups of $GL(n, K)$, to show how they give rise to K-loops, and to study some of their properties. In fact we can handle many classical groups.

B. GENERAL AND SPECIAL LINEAR GROUPS

(9.2) Theorem. *Let R be n-real, $K := R(i)$, and let*

$$G \in \{SL(n, R), GL(n, R), SL(n, K), GL(n, K)\}.$$

Then $G = L_G \Omega_G$, i.e., L_G is a transversal of G/Ω_G, and $(L_G, \circ)$ is a K-loop.

Proof. By (1.16) it suffices to show that for every $A \in G$, there exists $B \in L_G$ with $B^2 = AA^*$. By (1.13) all the eigenvalues of AA^* are squares in R. Hence (1.17) shows the result.   ∎

We'll determine the left inner mapping groups of our K-loops. This has been published in [65] by KIECHLE and KONRAD. It will be used to characterize those of our examples which have fixed point free left inner mappings. We note that by $PSO(2, R)$ we mean $SO(2, R)/\{\pm I_2\}$, despite the fact that $SO(2, R)$ is commutative.

(9.3) Theorem. *Let R be n-real, $K := R(i)$, and let*

$$G \in \{SL(n, R), GL(n, R), SL(n, K), GL(n, K)\}.$$

(1) $\mathcal{D}(L_G) \cong \begin{cases} PSO(n, R) & \text{if } G \in \{SL(n, R), GL(n, R)\} \\ PSU(n, K) & \text{if } G \in \{SL(n, K), GL(n, K)\}. \end{cases}$

(2) $\mathcal{Z}(L_G) = \mathcal{N}_r(L_G) = R^* I_n \cap L_G$

$$= \begin{cases} \{I_n\} & \text{if } G \in \{SL(n, R), SL(n, K)\} \\ \{rI_n; \, r > 0\} & \text{if } G \in \{GL(n, R), GL(n, K)\} \end{cases}.$$

(3) $(L_G, \circ)$ *has fixed point free left inner mappings if and only if $G = SL(2, R)$. This will work whenever R is a pythagorean field.*

(4) $PSL(2, R) = L_{SL(2,R)} \times_Q \mathcal{D}(L_{SL(2,R)})$ *is a specific group of characteristic 0.*

(5) L_G is uniquely 2-divisible if and only if R is euclidean.

Proof. (1) Let $D_G := \langle d_{A,B};\ A, B \in L_G \rangle$. From (2.8) or (9.1.2) one derives

$$D_G / C_{D_G}(L_G) \cong \mathcal{D}(L_G).$$

From (1.22) we infer $C_{D_G}(L_G) = D_G \cap I_n K^*$. So it suffices to show that

$$D_G = \begin{cases} \mathrm{SO}(n, R) & \text{if } G \in \{\mathrm{SL}(n, R), \mathrm{GL}(n, R)\} \\ \mathrm{SU}(n, K) & \text{if } G \in \{\mathrm{SL}(n, K), \mathrm{GL}(n, K)\}. \end{cases}$$

By (9.1.4) the inclusions "$\subseteq$" are immediate. It therefore remains to prove that

$$D_G \supseteq \begin{cases} \mathrm{SO}(n, R) & \text{if } G = \mathrm{SL}(n, R) \\ \mathrm{SU}(n, K) & \text{if } G = \mathrm{SL}(n, K). \end{cases}$$

This will be done in several steps. First we consider the case $G = \mathrm{SL}(2, R)$. We have

$$L_G = \left\{ \begin{pmatrix} \alpha & \beta \\ \beta & \gamma \end{pmatrix} \in R^{2 \times 2};\ \alpha > 0,\ \alpha\gamma - \beta^2 = 1 \right\},$$

$$\Omega_G = \mathrm{SO}(2, R) = \left\{ \begin{pmatrix} u & -v \\ v & u \end{pmatrix} \in R^{2 \times 2};\ u^2 + v^2 = 1 \right\}.$$

For two elements

$$A = \begin{pmatrix} \alpha & \beta \\ \beta & \gamma \end{pmatrix},\ B = \begin{pmatrix} \alpha' & \beta' \\ \beta' & \gamma' \end{pmatrix} \in L$$

let

$$\begin{pmatrix} a & b \\ c & d \end{pmatrix} := AB = \begin{pmatrix} \alpha\alpha' + \beta\beta' & \alpha\beta' + \gamma'\beta \\ \alpha'\beta + \gamma\beta' & \gamma\gamma' + \beta\beta' \end{pmatrix}$$

and

$$\Delta^2 := (a + d)^2 + (b - c)^2 = a^2 + b^2 + c^2 + d^2 + 2.$$

The last equality follows from $\det AB = ad - bc = 1$. Since R is pythagorean, we have $\Delta \in R$, and there is no loss to take $\Delta \geq 0$. Using this notation, we can give the explicit description

$$A \circ B = \frac{1}{\Delta} \begin{pmatrix} a^2 + b^2 + 1 & ac + bd \\ ac + bd & c^2 + d^2 + 1 \end{pmatrix} =: C$$

and

$$d_{A,B} = \frac{1}{\Delta} \begin{pmatrix} a+d & b-c \\ c-b & a+d \end{pmatrix} =: U.$$

Indeed, we'll prove $CU = AB$, i.e., $\Delta^2 AB = \Delta^2 CU =$

$$\begin{pmatrix} (a^2 + b^2 + 1)(a+d) & (a^2 + b^2 + 1)(b-c) \\ + (ac+bd)(c-b) & + (ac+bd)(a+d) \\ & \\ (ac+bd)(a+d) & (ac+bd)(b-c) \\ + (c^2 + d^2 + 1)(c-b) & + (c^2 + d^2 + 1)(a+d) \end{pmatrix}.$$

Using $ad - bc = 1$, we can compute

$$(a^2 + b^2 + 1)(a+d) + (ac+bd)(c-b) =$$
$$a(a^2 + b^2 + 1 + ad + c^2 - bc) + d(b^2 + 1 + bc - b^2) =$$
$$a(a^2 + b^2 + c^2 + d^2 + 2) = a\Delta^2,$$

hence the (1,1) entry behaves as claimed. In a very similar way, one can verify three more equalities for the other three entries to obtain the assertion.

Clearly $\det(U) = 1$, and so $U \in \Omega_G$. This also implies $\det(C) = 1$. By construction, C is symmetric and positive definite, hence it is an element of L_G. Now (9.2) shows $C = A \circ B$ and $U = d_{A,B}$.

Using this, all we need to do is to produce enough $d_{A,B}$'s to generate $SO(2, R)$. Put

$$A := \begin{pmatrix} \alpha & 0 \\ 0 & \alpha^{-1} \end{pmatrix}, \quad B := \begin{pmatrix} 1 & \beta \\ \beta & \gamma \end{pmatrix} \in L, \quad \alpha, \beta, \gamma \in R,$$

i.e., $\alpha > 0$ and $\gamma - \beta^2 = 1$. Then we compute

$$AB = \begin{pmatrix} \alpha & \alpha\beta \\ \alpha^{-1}\beta & \alpha^{-1}\gamma \end{pmatrix},$$

$$\Delta^2 = \alpha^2 + \alpha^2\beta^2 + \alpha^{-2}\beta^2 + \alpha^{-2}\gamma^2 + 2$$

and

$$d_{A,B} = \frac{1}{\Delta} \begin{pmatrix} \alpha + \alpha^{-1}\gamma & \beta(\alpha - \alpha^{-1}) \\ \beta(\alpha^{-1} - \alpha) & \alpha + \alpha^{-1}\gamma \end{pmatrix}. \tag{i}$$

Note that $(\alpha + \alpha^{-1}\gamma)\Delta^{-1} > 0$. We now prove

(a) *For all $c \in R$ there exist $\alpha, \beta, \gamma \in R$ with $\alpha > 0$, $\gamma - \beta^2 = 1$*
and $c = \dfrac{\beta(\alpha - \alpha^{-1})}{\alpha + \alpha^{-1}\gamma}$.

Proof of (a). Plugging in $1 + \beta^2$ for γ, the condition becomes

$$c = \frac{\beta(\alpha - \alpha^{-1})}{\alpha + \alpha^{-1}(1 + \beta^2)} \iff c(\alpha^2 + 1) + c\beta^2 = \beta(\alpha^2 - 1)$$

$$\iff c\beta^2 + (1 - \alpha^2)\beta + c(1 + \alpha^2) = 0.$$

We have to find $\alpha \in R$ with $\alpha > 0$ such that this quadratic equation in β has a solution. This is the case if and only if the discriminant D is a square in R, where

$$D := (1 - \alpha^2)^2 - 4c^2(1 + \alpha^2)$$
$$= \left(\alpha^2 - (2c^2 + 1)\right)^2 + 1 - \left((2c^2 + 1)^2 + 4c^2\right).$$

This is also true (though trivial) if $c = 0$. Since R is pythagorean,

$$y := \frac{1}{4}\left((2c^2 + 1)^2 + 4c^2\right)$$

is a square in R. Thus there is an $\alpha \in R$, $\alpha > 0$, such that $\alpha^2 := (2c^2 + 1) + y + 1$. The discriminant of our quadratic equation then becomes

$$D = \left((2c^2 + 1) + y + 1 - (2c^2 + 1)\right)^2 + 1 - 4y$$
$$= y^2 - 2y + 1 + 1 = (y - 1)^2 + 1.$$

This is a square in R, hence (a) is proved.

(b) *For every $U = \begin{pmatrix} u & v \\ -v & u \end{pmatrix} \in SO(2, R)$ with $u > 0$, there exist $A, B \in L$ such that $d_{A,B} = U$.*

Proof of (b). Let $c := \dfrac{v}{u}$. Then because of (a) and (i) there exist $A, B \in L$ with $d_{A,B} = \begin{pmatrix} u' & v' \\ -v' & u' \end{pmatrix}$ such that $c = \dfrac{v'}{u'}$. From

$$u^2 + v^2 = 1 = u'^2 + v'^2 \quad \text{and} \quad \frac{v}{u} = \frac{v'}{u'}$$

we obtain

$$1 = u^2 + \left(\frac{u}{u'}\right)^2 v'^2 = u^2 + \left(\frac{u}{u'}\right)^2 - \left(\frac{u}{u'}\right)^2 u'^2 = \left(\frac{u}{u'}\right)^2,$$

i.e., $u = u' (> 0)$. Hence $d_{A,B} = U$, and (b) is proved.

In particular, there are $A, B \in L$ with

$$d_{A,B} = \begin{pmatrix} \frac{1}{2}\sqrt{2} & \frac{1}{2}\sqrt{2} \\ -\frac{1}{2}\sqrt{2} & \frac{1}{2}\sqrt{2} \end{pmatrix},$$

hence

$$(d_{A,B})^2 = \begin{pmatrix} 0 & 1 \\ -1 & 0 \end{pmatrix} \quad \text{and} \quad (d_{A,B})^4 = -I_2.$$

From this one easily derives $SO(2, R) \subseteq D_{SL(2,R)}$.

We'll now take care of the case $G = SL(2, K)$. In fact, we can reduce to the previous case.

(c) *For every* $U \in SU(2, K)$ *there is* $T \in SU(2, K)$ *such that* $TUT^{-1} \in SO(2, R)$.

Proof of (c). Any $U \in SU(2, K)$ is of the form $U = \begin{pmatrix} u & v \\ -\bar{v} & \bar{u} \end{pmatrix}$.
Using $u\bar{u} + v\bar{v} = 1$ the characteristic polynomial of this matrix becomes $x^2 - (u + \bar{u})x + 1$, and the eigenvalues are

$$\begin{aligned}
\alpha &= \operatorname{Re}(u) + \sqrt{(u + \bar{u})^2/4 - u\bar{u} - v\bar{v}} \\
&= \operatorname{Re}(u) + \sqrt{(-1)(v\bar{v} - (u - \bar{u})^2/4)} \\
&= \operatorname{Re}(u) + i\sqrt{v\bar{v} + \operatorname{Im}(u)^2} = \alpha_1 + i\alpha_2
\end{aligned}$$

$\bar{\alpha} = \alpha_1 - i\alpha_2, \quad \alpha_1, \alpha_2 \in R$ (because R is pythagorean).

One easily verifies that the corresponding eigenvectors are

$$\mathbf{v} = \begin{pmatrix} \alpha_2 + \operatorname{Im}(u) \\ i\bar{v} \end{pmatrix}, \quad \mathbf{w} = \begin{pmatrix} iv(\alpha_2 - \operatorname{Im}(u)) \\ v\bar{v} \end{pmatrix},$$

where $\mathbf{v}$ belongs to α. A simple calculation shows that $\mathbf{v}$ and $\mathbf{w}$ are orthogonal.[1] Since R is pythagorean, there exist $r, s \in R$, $r, s > 0$, with

$$r^2 = \|\mathbf{v}\|^2 = \mathbf{v}^*\mathbf{v} = 2\alpha_2(\alpha_2 + \operatorname{Im}(u))$$

---

[1] An alternative approach would be the observation that U is a normal endomorphism.

and

$$s^2 = \|\mathbf{w}\|^2 = \mathbf{w}^*\mathbf{w} = 2\alpha_2 v\bar{v}(\alpha_2 - \operatorname{Im}(u)).$$

Put

$$S := \left(\frac{1}{r}\mathbf{v} \quad \frac{1}{s}\mathbf{w} \right) \in \mathrm{SU}(2, K),$$

then

$$S^{-1}US = \begin{pmatrix} \alpha & 0 \\ 0 & \bar{\alpha} \end{pmatrix} =: \Lambda.$$

An easy calculation shows that Λ is conjugate to

$$\begin{pmatrix} \alpha_1 & \alpha_2 \\ -\alpha_2 & \alpha_1 \end{pmatrix} \in \mathrm{SO}(2, R)$$

via the matrix $\frac{1}{\sqrt{2}}\begin{pmatrix} 1 & i \\ i & 1 \end{pmatrix} \in \mathrm{SU}(2, K)$. This proves (c).

Clearly, $D_{\mathrm{SL}(2,R)} \subseteq D_{\mathrm{SL}(2,K)}$. By (9.1.2) and (9.1.5) the normal subgroup generated by $D_{\mathrm{SL}(2,R)} = \mathrm{SO}(2, R)$ in $\mathrm{SU}(2, K)$ is contained in $D_{\mathrm{SL}(2,K)}$. Now, (c) shows that this group is equal to $\mathrm{SU}(2, K)$, hence $\mathrm{SU}(2, K) \subseteq D_{\mathrm{SL}(2,K)}$.

The general case can be reduced to the case $n = 2$ by using the embeddings

$$\Delta_k : \begin{cases} \mathrm{SL}(2, K) \to \mathrm{SL}(n, K) \\ A \mapsto \operatorname{diag}(I_{k-1}, A, I_{n-1-k}), \end{cases} \quad 1 \le k \le n - 1, \quad \text{(ii)}$$

introduced just before (1.19). Now (1.19) and (2.7.6) imply for all $A, B \in \mathcal{H}(n, K)$

$$\Delta_k(d_{A,B}) = d_{\Delta_k(A),\Delta_k(B)}, \quad \text{hence} \quad \Delta_k\big(\mathrm{SU}(2, K)\big) \subseteq D_{\mathrm{SL}(n,K)}$$

with what we have proved above. (1.20) shows the result for the case $G = \mathrm{SL}(n, K)$. The case $G = \mathrm{SL}(n, R)$ runs exactly the same.

(2) Let $A \in L_G$, $r \in R$ with $rI_n \in L_G$, then $r > 0$, and $rI_nA = rA \in L_G$. Therefore, $d_{rI_n,A} = I_n$, and $rI_n \in \mathcal{N}_\ell = \mathcal{N}_m = \mathcal{Z} \subseteq \mathcal{N}_r$ by (9.1.2), (5.1.2) and (5.10.2). By (5.4.2), we're left to showing that $\mathcal{N}_r = \operatorname{Fix}\big(\mathcal{D}(L_G)\big) \subseteq \{rI_n; \ r > 0\}$.

Noting that $\det(rI_n) = r^n = 1 \iff r = 1$, it suffices to do the cases $G = \mathrm{GL}(n, R)$ and $G = \mathrm{GL}(n, K)$. Let $A \in \mathcal{N}_r$. By (1.12) there exists $\omega \in \mathrm{SO}(n, R)$, or $\omega \in \mathrm{SU}(n, K)$, respectively, such that $\widehat{\omega}(A) = \mathrm{diag}(\alpha_1, \ldots, \alpha_n)$, with $\alpha_1, \ldots, \alpha_n \in R$. From (1) we know $\widehat{\omega} \in \mathcal{D}(L_G)$, hence $A = \widehat{\omega}(A) = \mathrm{diag}(\alpha_1, \ldots, \alpha_n)$.

A simple calculation shows that for all $\beta_1, \beta_2 \in K$

$$\begin{pmatrix} 0 & 1 \\ -1 & 0 \end{pmatrix} \mathrm{diag}(\beta_1, \beta_2) \begin{pmatrix} 0 & 1 \\ -1 & 0 \end{pmatrix}^{-1} = \mathrm{diag}(\beta_2, \beta_1). \qquad \text{(iii)}$$

Using the embeddings $\Delta_k : \mathrm{GL}(2, K) \to \mathrm{GL}(n, K)$ from (1.19) (see also (ii)) one easily derives $\alpha_k = \alpha_{k+1}$ for all $k \in \{1, \ldots, n-1\}$, and $A = \alpha_1 I_n$. Clearly, $\alpha_1 > 0$.

(3) Assume first that $\mathcal{D}(L_G)$ acts fixed point free on L_G. If $A := \mathrm{diag}(\lambda, \lambda, \ldots) \in L_G$ for some $\lambda > 1$, then (iii) shows $\omega A \omega^{-1} = A$, for

$$\omega := \mathrm{diag}\left(\begin{pmatrix} 0 & 1 \\ -1 & 0 \end{pmatrix}, I_{n-2} \right) \in \mathrm{SO}(n, R) \subseteq \Omega_G.$$

Moreover, $\widehat{\omega} \neq 1$, by (1.22). By (1), $\widehat{\omega} \in \mathcal{D}(L_G)$, and $\widehat{\omega}$ is not fixed point free, a contradiction.

Therefore we must have $n = 2$ and $G = \mathrm{SL}(2, .)$. [2] The case $G = \mathrm{SL}(2, K)$ must be excluded, because $\omega := \mathrm{diag}(i, -i) \in \Omega_G$ and $\widehat{\omega}$ has fixed point $\mathrm{diag}(\lambda, \lambda^{-1}) \in L_G$ for every positive $\lambda \in R$, but $\widehat{\omega} \neq 1$ by (1.22). Also, $\widehat{\omega} \in \mathcal{D}(L_G)$ by (1).

Now assume $G = \mathrm{SL}(2, R)$, then $\Omega_G = \mathrm{SO}(2, R)$. Let $A \in L_G$, $A \neq I_2$, and choose $\omega \in \Omega_G$ with $\omega A \omega^{-1} = A$. By (1.12) there is $S \in \mathrm{SO}(2, R)$ such that

$$\Lambda := SAS^{-1} = \mathrm{diag}(\lambda, \lambda^{-1})$$

is diagonal. The second eigenvalue of Λ has to be λ^{-1}, because $\det \Lambda = 1$. For $\omega' := S\omega S^{-1}$ we have $\omega' \Lambda \omega'^{-1} = \Lambda$ and $\omega' \in \Omega_G$. Since $\lambda > 0$ and $\lambda \neq 1$, we have $\lambda \neq \lambda^{-1}$. From (1.18) we conclude that $\omega' = \mathrm{diag}(\alpha_1, \alpha_2)$ and $|\alpha_1| = |\alpha_2| = 1$. Since $\det \omega' = 1$, we have $\alpha_1 = \alpha_2 = \pm 1$. Thus $\widehat{\omega'} = 1$, and so $\widehat{\omega} = 1$.

---

[2] $G = \mathrm{SL}(n, .)$ also follows from (2) and (5.10.2).

(4) By (3) and (7.19), all we need to show is that every element $A \in L_{\mathrm{SL}(2,R)}^{\#}$ has infinite order. Because of (9.1.4) we can look at A is G. The eigenvalues of A are all positive elements of R, not all equal to 1. Such elements have infinite order, thus so has A.

(5) Assume that L_G is 2-divisible, and let $a \in R$, $a > 0$. Then the matrix $A := \mathrm{diag}(a, a^{-1}, 1, \ldots, 1) \in L_G$ in all four cases. Since A has a square root in L_G, then a must have a square root in R. Therefore R is euclidean.

Let R be euclidean, and choose $A \in L_G$. By (1.12) the eigenvalues of A are positive elements of R, hence squares. Now (1.17) shows that A has a square root in L_G. This is unique by (1.14).   ∎

In (1), we have actually proved a little more. This will be useful later.

(9.4) *Let R be n-real, $K := R(i)$, then*

$$
\begin{aligned}
D_G :&= \langle d_{A,B};\ A, B \in L_G \rangle \\
&= \begin{cases} \mathrm{SO}(n, R) & \text{if } G \in \{\mathrm{SL}(n, R), \mathrm{GL}(n, R)\} \\ \mathrm{SU}(n, K) & \text{if } G \in \{\mathrm{SL}(n, K), \mathrm{GL}(n, K)\}. \end{cases}
\end{aligned}
$$
■

For the next example, we'll need some preparation.

(9.5) *Let K be an arbitrary field and $A \in \mathrm{GL}(2, K)$ such that $\det A = \alpha^2$ is a square in K, i.e., $\alpha \in K$. Put $\beta := \mathrm{tr}\,A$ and assume $\sqrt{2\alpha + \beta} \in K$. For*

$$
B := \frac{1}{\sqrt{2\alpha + \beta}}(\alpha I_2 + A)
$$

we have

(1) $B^2 = A$.

(2) *If $K = R(i)$ with an ordered field R, and if A is hermitian with $\alpha, \sqrt{2\alpha + \beta} \in R$, then B is hermitian.*

(3) *If, in addition to the assumptions of (2), A is positive definite, and $\alpha, \sqrt{2\alpha + \beta} > 0$, then B is positive definite.*[3]

---

[3] Notice that "positive definite" implies $\alpha, \sqrt{2\alpha + \beta} \in R$.

(4) *With assumptions as in (3), the conditions B hermitian, positive definite with $B^2 = A$, uniquely determine B. In this case we write $\sqrt{A} := B$.*

The *Proof* of the first part comes easily from the Cayley-Hamilton-Theorem. It is also easy to see that B is hermitian in case (2).

(3) To see that B is positive definite, simply compute $\mathbf{v}^* B \mathbf{v}$ for general $\mathbf{v} \in R(i)^2$.

(4) follows directly from (1.14). ∎

For our example, we'll assume that R is an arbitrary, ordered field. Write R^{real} for a (fixed) real closure of R.

Let $K = R(i)$ as before, $n = 2$, and let $GL'(2, K) := \{A \in GL(2, K); \det A \in R\}$, the invertible 2×2-matrices over K with "real" determinant. Consider

$$G := \bigcup_{d \in R,\, d > 0} \sqrt{d}\, GL'(2, K) \subseteq GL\big(2, R^{\text{real}}(i)\big),$$

where $\sqrt{d}$ is chosen positive in R^{real}. Clearly, G is a subgroup of $GL\big(2, R^{\text{real}}(i)\big)$.

(9.6) For all $A \in L_G$ we have $AA^* \in \kappa(L_G)$, hence L_G is a transversal of G/Ω_G and $(L_G, \circ)$ is a K-loop.

Proof. We have $A = \sqrt{d}B$ for some $d > 0$ and some $B \in GL'(2, K)$. Thus $AA^* = dBB^*$ and $\det(dBB^*) = d^2(\det B)^2$ is a square in R. Also, AA^* is hermitian, positive definite, so by (9.5) $C := \sqrt{AA^*} \in L_G$ satisfies $\kappa(C) = C^2 = AA^*$.

The remaining assertions follow from the preceding paragraph, (1.16), and (9.1). ∎

Remarks. **1.** The above example is isomorphic to the example described by IM [48] via the map κ. More precisely: Let $\widetilde{\mathcal{H}}$ be the set of positive definite 2×2-matrices over K with square determinant, i.e., for all $A \in \widetilde{\mathcal{H}}$ there is $\alpha \in R$ such that $\det A = \alpha^2$. The map $\kappa : L_G \to \widetilde{\mathcal{H}}; X \mapsto X^2$ is bijective and $(\widetilde{\mathcal{H}}, +)$ becomes a K-loop by (see also (6.14))

$$A \oplus B := \kappa\big(\kappa^{-1}(A) \circ \kappa^{-1}(B)\big) = \kappa^{-1}(A)B\kappa^{-1}(B).$$

This is IM's example! To see that κ is actually well-defined and bijective, note first that $\kappa(X) \overset{!}{=} XX^*$ does have square determinant. From (1.14) we know that κ is injective. Finally, for $B \in \widetilde{\mathcal{H}}$ (9.5) gives $\kappa^{-1}(B) = \sqrt{B} \in L_G$.

2. As we have indicated, IM's construction uses the method of (6.14). In fact some of the examples above and the one of (9.9) below have been constructed this way.

3. KIKKAWA was probably the first to observe that the method of (6.14) could be applied to sets of positive definite symmetric and hermitian matrices over the reals, to give what we call left power alternative Kikkawa loops, [67; Ex. 1.5]. He didn't mention the Bol identity, though.

C. LINEAR GROUPS OVER THE QUATERNIONS

Throughout this subsection, we assume that R is n-real. We begin with a construction which will also be useful, when we deal with symplectic, pseudo-orthogonal and pseudo-unitary groups in the following subsections.

Let τ be an involutory automorphism of $\mathrm{GL}(n, K)$ which commutes with $*$, i.e., $(A^*)^\tau = (A^\tau)^*$ for all $A \in \mathrm{GL}(n, K)$. Moreover, we assume that τ induces a map $R \setminus \{0\} \to R \setminus \{0\}$, also denoted by τ, such that

$$\mathrm{diag}(\alpha_1, \ldots, \alpha_n)^\tau = \mathrm{diag}(\alpha_1^\tau, \ldots, \alpha_n^\tau), \quad \text{and} \quad \alpha > 0 \iff \alpha^\tau > 0,$$

for all $\alpha, \alpha_1, \ldots, \alpha_n \in R \setminus \{0\}$.

There are three examples for τ, which we shall be interested in:

$$A \mapsto \bar{A}, \quad A \mapsto (A^{-1})^{\mathrm{T}} \quad \text{and} \quad A \mapsto A^{-*} := (A^{-1})^*.$$

Furthermore, let $J \in \mathrm{GL}(n, K)$ be such that $J^* = \lambda J^{-1}$ for some $\lambda \in R$. We'll consider

$$G_{\tau, J}^{(n)} := \{A \in \mathrm{GL}(n, K);\ A^\tau = JAJ^{-1}\},$$

which is clearly a subgroup of $GL(n, K)$.

(9.7) Theorem. *For $A \in G := G_{\tau,J}^{(n)}$, we have*

(1) $A^* \in G$ *and* $\sqrt{\alpha}^\tau = \sqrt{\alpha^\tau}$, α *a square in* $R \setminus \{0\}$;

(2) *There exists $B \in L_G$ with $AA^* = B^2$, hence L_G is a transversal of G/Ω_G and $(L_G, \circ)$ becomes a K-loop as in (9.1). This holds verbatim if $G = G_{\tau,J}^{(n)} \cap GL(n, R)$.*

Proof. (1) We have $A^\tau = JAJ^{-1}$, hence

$$(A^*)^\tau = (A^\tau)^* = (J^{-1})^* A^* J^* = (J^*)^{-1} A^* J^*$$
$$= \lambda \lambda^{-1} J A^* J^{-1} = J A^* J^{-1}.$$

For the second assertion, it suffices to note that $\tau : R \setminus \{0\} \to R \setminus \{0\}$ is multiplicative, and $\sqrt{\alpha}^\tau > 0$.

(2) Observe that B does exist in $\mathcal{H}(n, K) = L_{GL(n,K)}$ by (9.2.1) and (1.16), and is unique by (1.14). Thus it is enough to show that $B \in G$.

Let $\alpha_1, \ldots, \alpha_n$ be the eigenvalues of AA^*. By the spectral theorem (1.12) and the required properties of τ, the eigenvalues of $(AA^*)^\tau$ are $\alpha_1^\tau, \ldots, \alpha_n^\tau$. By (1.13) the α_k's are squares in R. By (1) so are the α_k^τ's. Therefore we can apply (1.17) to the matrices AA^* and $(AA^*)^\tau$. According to (1.17) there exists a polynomial $f \in R[x]$ with

$$B = f(AA^*), \quad f((AA^*)^\tau) \in \mathcal{H}(n, K)$$

and

$$\left(f((AA^*)^\tau)\right)^2 = (AA^*)^\tau.$$

Since also $(B^\tau)^2 = (B^2)^\tau = (AA^*)^\tau$, we must have $f((AA^*)^\tau) = B^\tau$ by (1.14). Therefore, using (1), we can compute

$$JBJ^{-1} = Jf(AA^*)J^{-1} = f(JAA^*J^{-1}) = f((AA^*)^\tau) = B^\tau,$$

hence $B \in G$. Since f is a polynomial over R, we can conclude that $B \in GL(n, R)$ if $AA^* \in GL(n, R)$. ∎

For a first application let $H := R(i, j)$ be the quaternions over R, where i, j satisfy the usual identities (see [43]). Note that H is a non-commutative skewfield.

Let $\tau : \mathrm{GL}(2n, K) \to \mathrm{GL}(2n, K); A \mapsto \bar{A}$ and

$$J := \mathrm{diag}\left(\begin{pmatrix} 0 & -1 \\ 1 & 0 \end{pmatrix}, \ldots, \begin{pmatrix} 0 & -1 \\ 1 & 0 \end{pmatrix} \right) \in \mathrm{GL}(2n, R)$$

By [43; I.2 Satz 17, p. 33] we have

(9.8) $\mathrm{GL}(n, H) \cong G_{\tau, J}^{(2n)}$, *where quaternions are identified with complex 2×2-matrices in the canonical way.* ∎

Using the above identification, we let $\mathrm{SL}(n, H) := \mathrm{SL}(2n, K) \cap \mathrm{GL}(n, H)$, and we obtain

(9.9) *Let $G \in \{\mathrm{SL}(n, H), \mathrm{GL}(n, H)\}$. Then L_G is a transversal of G/Ω_G and is therefore a K-loop.*

Proof. Obviously, τ commutes with $*$ and $J^* = J^{\mathrm{T}} = J^{-1}$. Thus for $\mathrm{GL}(n, H)$ the assertion is immediate from (9.7.2). If $G = \mathrm{SL}(n, H)$, let $A \in G$. Then $B \in L_{\mathrm{GL}(n, H)}$ with $B^2 = AA^*$ necessarily has determinant 1, hence the result. ∎

Remark. In the case $n = 2$ this example has been found by KON-RAD [76], using a different approach. KONRAD has used euclidean fields, while we need 4-real fields. By (1.7) there exist 4-real fields which are not euclidean. It seems open, whether every euclidean field is 4-real, but notice the second statement in (1.7).

D. SYMPLECTIC GROUPS

To define the symplectic groups take

$$\tau : \mathrm{GL}(2n, K) \to \mathrm{GL}(2n, K); A \mapsto (A^{-1})^{\mathrm{T}}$$

and

$$J := \begin{pmatrix} 0 & I_n \\ -I_n & 0 \end{pmatrix} \in R^{2n \times 2n}.$$

The groups $\mathrm{Sp}(2n, K) := \{A \in \mathrm{GL}(2n, K); A^{\mathrm{T}} J A = J\} = G_{\tau, J}^{(2n)}$ and $\mathrm{Sp}(2n, R) := \mathrm{Sp}(2n, K) \cap \mathrm{GL}(2n, R)$ are called the *symplectic groups* over K, R, respectively (see [47; II.9]). By [47; II.9.19,

p. 224] we have $\mathrm{Sp}(2n, K) \subseteq \mathrm{SL}(2n, K)$. We will not really need this fact, but it explains why there is no point in intersecting $\mathrm{Sp}(2n, K)$ with $\mathrm{SL}(2n, K)$. Recall that $\mathrm{Sp}(2, K) = \mathrm{SL}(2, K)$ [47; II.9.12, p. 219], so the interesting cases start with $n \geq 2$. A direct application of (9.7.2) gives

(9.10) Theorem. If $G \in \{\mathrm{Sp}(2n, R), \mathrm{Sp}(2n, K)\}$, then L_G is a transversal of G/Ω_G and is therefore a K-loop.   ∎

E. PSEUDO-ORTHOGONAL AND PSEUDO-UNITARY GROUPS

Our third application of the construction in C will be to pseudo-orthogonal and pseudo-unitary groups over R and K, respectively. We continue to assume that R is n-real.

Let $\tau = {}^{-*}$, i.e., $A^\tau = (A^{-1})^* = (A^*)^{-1}$, let $p, q \in \mathbf{N}$ be such that $p + q = n$, and put

$$J := \mathrm{diag}(\underbrace{1, \ldots, 1}_{p}, \underbrace{-1, \ldots, -1}_{q}) = \mathrm{diag}(I_p, -I_q) \in \mathrm{GL}(n, R).$$

Then $(\mathbf{v}, \mathbf{w}) \mapsto \mathbf{v}^* J \mathbf{w}$ defines a hermitian form on K^n, or a symmetric form on R^n, respectively. The corresponding isometry groups are the pseudo-unitary, pseudo-orthogonal groups, respectively. More precisely,

$$\mathrm{U}(p, q) := G_{\tau, J}^{(n)} = \{A \in \mathrm{GL}(n, K); \ A^* J A = J\},$$
$$\mathrm{O}(p, q) := \mathrm{U}(p, q) \cap \mathrm{GL}(n, R).$$

Note that the fields K, R, respectively, are tacitly understood, and thus suppressed from the notation. We also write briefly $\mathrm{U}(n), \mathrm{O}(n)$ for $\mathrm{U}(n, K), \mathrm{O}(n, R)$, respectively.

Directly from (9.7.2) we get

(9.11) Theorem. Let $G \in \{\mathrm{O}(p, q), \mathrm{U}(p, q)\}$. Then L_G is a transversal of G/Ω_G and is therefore a K-loop.   ∎

We shall now take a closer look at the so constructed K-loops. We begin with a useful characterization of $\mathrm{U}(p, q)$.

(9.12) Write $A \in GL(n, K)$ as $A = \begin{pmatrix} S & X \\ Y & T \end{pmatrix}$, with $S \in K^{p \times p}$, $X \in K^{p \times q}$, $Y \in K^{q \times p}$, $T \in K^{q \times q}$, then

(1) $A \in U(p, q)$ if and only if

$$S^*S = I_p + Y^*Y, \quad T^*T = I_q + X^*X, \quad S^*X = Y^*T.$$

(2) If $A \in L_{U(p,q)}$, then S, T are positive definite, hermitian, and $Y = X^*$.

(3) Given $X \in K^{p \times q}$, there exist unique positive definite, hermitian $S \in K^{p \times p}$, $T \in K^{q \times q}$ such that

$$S^2 = I_p + XX^* \quad \text{and} \quad T^2 = I_q + X^*X.$$

Moreover,

$$A := \begin{pmatrix} S & X \\ X^* & T \end{pmatrix} \in L_{U(p,q)}.$$

In particular, A is uniquely determined by X.

Proof. (1) is easy to see (cf. [43; p. 52]).

(2) If A is hermitian, then $S = S^*$, $T = T^*$ and $Y = X^*$. It remains to show that S, T are positive definite. This is easy to see, and can be left to the reader.

(3) By (1) S and T must satisfy the given conditions:

$$S^2 = I_p + XX^*, \quad T^2 = I_q + X^*X \quad \text{and also} \quad SX = XT.$$

By (1.13) we know that the eigenvalues of XX^* and X^*X are squares in R. Since R is pythagorean, the eigenvalues of $I_p + XX^*$ and $I_q + X^*X$ are also squares in R. Noting that R is p- and q-real, (1.17) shows that $I_p + XX^*$ and $I_q + X^*X$ do have square roots in $\mathcal{H}(p, K)$, $\mathcal{H}(q, K)$, respectively. These square roots are unique by (1.14). Therefore S and T are uniquely determined, given X.

Next, we prove that $SX = XT$ (to make sure that A is hermitian). Using (1.17) again, there exists a polynomial $f \in R[x]$ such that $S = f(S^2)$ and $T = f(T^2)$. Therefore, it suffices to show that $S^2X = XT^2$, but this is clear from the equations for S^2 and T^2.

Finally, we have to show that A is positive definite. We have

$$A = F \operatorname{diag}(S, T^{-1}) F^*, \quad \text{where} \quad F = \begin{pmatrix} I_p & 0 \\ X^* S^{-1} & I_q \end{pmatrix},$$

because $X^* S^{-1} X + T^{-1} = X^* X T^{-1} + T^{-1} = (X^* X + I_q) T^{-1} = T$. Since $\operatorname{diag}(S, T^{-1})$ is positive definite by construction of S, T, the result follows. ∎

We note in passing that part (2) of the preceding lemma holds for arbitrary positive definite, hermitian matrices.

$A \in U(p, q)$ will frequently be written as a block matrix, with appropriate sizes as in (9.12) tacitly understood.

We can now determine the structure of Ω_G. To get handy notation, write

$$U(p) \times U(q) = \{\operatorname{diag}(S, T); \ S \in U(p, K), \ T \in U(q, K)\} \subseteq U(p, q),$$

and analogously for $O(p) \times O(q) \subseteq O(p, q)$. Notice that $U(p) \times U(q)$ really is isomorphic to the direct product of $U(p)$ and $U(q)$.

(9.13) Theorem. *Let $G \in \{O(p, q), U(p, q)\}$. Then*

$$\Omega_G = \begin{cases} U(p) \times U(q) & \text{for } G = U(p, q); \\ O(p) \times O(q) & \text{for } G = O(p, q). \end{cases}$$

The *Proof* will be done for $G = U(p, q)$ only. The other case is exactly the same.

Obviously, $U(p) \times U(q) \subseteq \Omega_G$ (see also (9.12)).

Let $e_1, \ldots, e_n$ be the standard basis of K^n. Observe that $\omega \in \Omega_G$ is isometric with respect to the hermitian forms

$$(\mathbf{v}, \mathbf{w}) \mapsto \mathbf{v}^* \mathbf{w} \quad \text{and} \quad (\mathbf{v}, \mathbf{w}) \mapsto \mathbf{v}^* J \mathbf{w}.$$

Let $k \in \{1, \ldots, p\}$, and $\omega e_k = (v_1, \ldots, v_n)$. Then $e_k^* e_k = 1$ and $e_k^* J e_k = 1$, therefore

$$(\omega e_k)^* (\omega e_k) = |v_1|^2 + \ldots + |v_n|^2 = 1$$

and

$$(\omega \mathbf{e}_k)^* J (\omega \mathbf{e}_k) = |v_1|^2 + \ldots + |v_p|^2 - |v_{p+1}|^2 - \ldots - |v_n|^2 = 1.$$

Subtracting these two equations yields

$$2(|v_{p+1}|^2 + \ldots + |v_n|^2) = 0, \quad \text{hence} \quad v_{p+1} = \ldots = v_n = 0.$$

If $k \in \{p+1, \ldots, n\}$, then $\mathbf{e}_k^* \mathbf{e}_k = 1$ and $\mathbf{e}_k^* J \mathbf{e}_k = -1$. A very similar line of reasoning shows $v_1 = \ldots = v_p = 0$. This proves that

$$\omega = \mathrm{diag}(S, T) \quad \text{with} \quad S \in \mathrm{GL}(p, K), \, T \in \mathrm{GL}(q, K).$$

By (9.12.1), S and T are unitary. $\blacksquare$

The next lemma shows that we could have started just as well with $\mathrm{SU}(p, q)$, $\mathrm{SO}(p, q)$ the set of elements of $\mathrm{U}(p, q)$, $\mathrm{O}(p, q)$, respectively, of determinant 1. However, for the sake of generality we chose the present approach. It also gives a new proof for $\det d_{A,B} = 1$ (see (9.1.4)) in the present special situation.

(9.14) For $A \in L_{\mathrm{U}(p,q)}$, we have $\det A = 1$.

Proof. The defining equation for elements of $\mathrm{U}(p, q)$ gives

$$\overline{(\det A)}(\det A) = 1.$$

Since A, being positive definite, has positive determinant, we conclude that $\det A = 1$. $\blacksquare$

It turns out that some of our "new" examples are already known. Here we can take advantage of the preceding lemma.

(9.15) *Every element of* $\mathrm{SU}(1, 1)$ *is of the form*

$$\begin{pmatrix} a & b \\ \bar{b} & \bar{a} \end{pmatrix}, \quad a, b \in K, \quad a\bar{a} - b\bar{b} = 1.$$

The map

$$\phi : \begin{cases} \mathrm{SU}(1, 1) \to \mathrm{SL}(2, R) \\ \begin{pmatrix} a & b \\ \bar{b} & \bar{a} \end{pmatrix} \mapsto \begin{pmatrix} a_1 + b_1 & -a_2 + b_2 \\ a_2 + b_2 & a_1 - b_1 \end{pmatrix} \end{cases}$$

where $a = a_1 + a_2 i$, $b = b_1 + b_2 i$, is an isomorphism, with

$$\phi(L_{SU(1,1)}) = L_{SL(2,R)} \quad \text{and} \quad \phi(\Omega_{SU(1,1)}) = \Omega_{SL(2,R)}.$$

Hence the loops $L_{SU(1,1)} = L_{U(1,1)}$ and $L_{SL(2,R)}$ are isomorphic via ϕ.

Proof. Let $A = \begin{pmatrix} a & b \\ c & d \end{pmatrix} \in SU(1,1)$. By (9.12.1), and $\det A = 1$, we have

$$\bar{a}b = \bar{c}d, \quad a\bar{a} - c\bar{c} = 1 \quad \text{and} \quad ad - bc = 1,$$

therefore

$$a\bar{a}b = a\bar{c}d = \bar{c}(1 + bc) = \bar{c} + bc\bar{c}$$
$$\implies b = b(a\bar{a} - c\bar{c}) = \bar{c} \quad \text{and} \quad \bar{a} = d.$$

This shows that the elements of $SU(1,1)$ are of the given form. Furthermore, the R-subspace $U := \{(z, \bar{z})^T; z \in K\} \subseteq K^2$ is invariant under the action of $SU(1,1)$. Choosing the basis $(1,1)^T$, $(i, -i)^T$, of U, gives the homomorphism ϕ. Let

$$A = \begin{pmatrix} a & b \\ \bar{b} & \bar{a} \end{pmatrix} \in SU(1,1),$$

then $\det \phi(A) = a_1^2 + a_2^2 - b_1^2 - b_2^2 = a\bar{a} - b\bar{b} = 1$, hence $\phi(A) \in SL(2,R)$.

An easy calculation shows that $\phi(A) = I_2$ implies $a = 1$, $b = 0$, so the kernel of ϕ is trivial, and ϕ is injective.

A is positive definite, hermitian if and only if $a = a_1 = \sqrt{1 + b\bar{b}}$, so in particular, $a_1 > 0$, $a_2 = 0$, if and only if $\phi(A)$ is positive definite, symmetric.

By (9.13) $A \in \Omega_{SU(1,1)}$ if and only if $b = 0$ if and only if $\phi(A) \in SO(2) = \Omega_{SL(2,R)}$.

Using the polar decomposition (9.2) for $SL(2,R)$ gives an easy proof for the surjectivity of ϕ. (It's not hard to do this directly, either.) ∎

Next we consider the special case $O(p,1)$. The connection with relativity will be explored in §10.

(9.16) *Let* $A, B \in L_{O(p,1)}$, *then* $d_{A,B} = \mathrm{diag}(U, 1)$ *for* $U \in \mathrm{SO}(p)$.

Proof. Using (9.12), we can write

$$A = \begin{pmatrix} S_1 & X_1 \\ X_1^* & t_1 \end{pmatrix} \quad \text{and} \quad B = \begin{pmatrix} S_2 & X_2 \\ X_2^* & t_2 \end{pmatrix},$$

where $t_k = \sqrt{1 + X_k^* X_k} > 0$, $(k = 1, 2)$. (Note that here $X^* = X^{\mathrm{T}}$.) The (n, n)-entry of AB is $r := X_1^* X_2 + t_1 t_2$.

By the Schwarz inequality (1.8) we get the estimate

$$|X_1^* X_2| \leq \|X_1\| \|X_2\| < \sqrt{1 + X_1^* X_1} \sqrt{1 + X_2^* X_2} = t_1 t_2.$$

This implies that $r > 0$. The (n, n)-entry t_3 of $A \circ B$ is positive, too. Since $AB = (A \circ B) d_{A,B}$, and $d_{A,B} = \mathrm{diag}(U, e)$, we can conclude that $r = t_3 e$, and e is positive, as well. But $e \in \mathrm{O}(1) = \{\pm 1\}$ (see (9.13)), hence $e = 1$. Finally, by (9.1.4) or (9.14), $\det U = \det d_{A,B} = 1$. So $U \in \mathrm{SO}(p)$. ∎

Since $\mathrm{SO}(1) = \{1\}$, we obtain an immediate corollary.

(9.17) $L_{O(1,1)}$ *is an abelian group.* ∎

Remarks. 1. If R is the field of real numbers, then (9.16) holds in much more generality. Indeed, the connected component of I_n in the (Lie-)group $\mathrm{O}(p, q)$ is a subgroup of index four, which is therefore strictly contained in $\mathrm{SO}(p, q)$ (cf. [44; IX.4.4 (b), p. 346]). It is characterized by the condition that $\det S, \det T > 0$, where elements are written in the form of (9.12). This implies that $d_{A,B} \in \mathrm{SO}(p) \times \mathrm{SO}(q)$ for all $A, B \in L_{O(p,q)}$.

2. One can easily show that $L_{O(1,1)}$ is isomorphic to the group of positive elements of R, applying the coordinate transformation given by $\omega := \dfrac{\sqrt{2}}{2} \begin{pmatrix} 1 & 1 \\ -1 & 1 \end{pmatrix}$ to the representation of $L_{O(1,1)}$ derived from (9.12.3).

(9.18) *There exist homomorphisms*

$$\mathrm{SL}(2, R) \to \mathrm{SO}(2, 1) \quad \text{and,} \quad \mathrm{SL}(2, K) \to \mathrm{SO}(3, 1)$$

which induce injective maps

$$L_{\mathrm{SL}(2,R)} \to L_{\mathrm{SO}(2,1)} \quad \text{and,} \quad L_{\mathrm{SL}(2,K)} \to L_{\mathrm{SO}(3,1)},$$

respectively. These maps are monomorphisms of the K-loops. If R is euclidean, then they are isomorphisms.

Proof. Let

$$\mathcal{H} := \left\{ \begin{pmatrix} \alpha & b \\ \bar{b} & \gamma \end{pmatrix} ; \ \alpha, \gamma \in R, \ b = b_1 + ib_2 \in K \right\},$$

and

$$\mathcal{S} := \left\{ \begin{pmatrix} \alpha & \beta \\ \beta & \gamma \end{pmatrix} \in \mathcal{H}; \ \beta \in R \right\},$$

be the set of hermitian, and symmetric 2×2-matrices over K, R, respectively. $\mathcal{H}$ and $\mathcal{S}$ are vector spaces over R of dimension 4, 3, respectively. Indeed,

$$\begin{pmatrix} 0 & i \\ -i & 0 \end{pmatrix}, \ \begin{pmatrix} 0 & 1 \\ 1 & 0 \end{pmatrix}, \ \begin{pmatrix} 1 & 0 \\ 0 & -1 \end{pmatrix}, \ \begin{pmatrix} 1 & 0 \\ 0 & 1 \end{pmatrix}$$

is a basis of $\mathcal{H}$. Deleting the first matrix gives a basis of $\mathcal{S}$.

The determinant D is a quadratic form on $\mathcal{H}$, given by the matrix $-\operatorname{diag}(I_3, -1)$, with respect to the above (ordered) basis. Therefore, the isometry group of $\mathcal{H}$ is isomorphic to $O(3, 1)$. Analogously, the isometry group of $\mathcal{S}$ is isomorphic to $O(2, 1)$.

Let tr be the trace on $K^{2 \times 2}$. Then $T(X, Y) := \frac{1}{2} \operatorname{tr}(XY)$ is a symmetric bilinear form on $\mathcal{H}$, which has Gram-matrix I_4, with respect to our basis. Again an analogous statement holds for $\mathcal{S}$.

Now $SL(2, R)$, $SL(2, K)$, respectively, act on $\mathcal{S}$, $\mathcal{H}$, respectively, by $A^\diamond(X) := AXA^*$, where $A \in SL(2, K)$. This action gives isometries of $\mathcal{S}$, $\mathcal{H}$ with respect to D, hence the desired homomorphisms. More specific: if $\phi(A^\diamond)$ is the matrix representation of $A^\diamond$ with respect to the above (ordered) basis of $\mathcal{H}$, then the map $\phi : SL(2, K) \to O(3, 1)$ is the desired homomorphism in the first case. The second case is analogous.

We're going to prove

(a) *The kernel of ϕ is $\{\pm I_2\}$, in both cases.*

Indeed, let $A \in SL(2, K)$ with $\phi(A) = I_4$. Then $AI_2A^* = I_2$ and $A \operatorname{diag}(1, -1)A^* = \operatorname{diag}(1, -1)$, hence $A \in SU(2, K) \cap SU(1, 1)$. By

(9.13) $A = \text{diag}(a, \bar{a})$, $a\bar{a} = 1$. Moreover,

$$A \begin{pmatrix} 0 & 1 \\ 1 & 0 \end{pmatrix} A^* = \begin{pmatrix} 0 & a^2 \\ \bar{a}^2 & 0 \end{pmatrix} \overset{!}{=} \begin{pmatrix} 0 & 1 \\ 1 & 0 \end{pmatrix} \implies a = \pm 1.$$

This implies (a).

Our next claim is that $U \in \text{SU}(2, K)$ is mapped to an isometry with respect to T, indeed for $X, Y \in \mathcal{H}$

$$T(U^\diamond(X), U^\diamond(Y)) = \frac{1}{2} \text{tr}(UXU^{-1}UYU^{-1}) = T(X, Y).$$

using that $U^* = U^{-1}$. Therefore,

$$\phi(\Omega_{\text{SL}(2,R)}) \subseteq \Omega_{\text{O}(2,1)} \quad \text{and,} \quad \phi(\Omega_{\text{SL}(2,K)}) \subseteq \Omega_{\text{O}(3,1)},$$

respectively. We now claim that a hermitian matrix A in $\text{SL}(2, K)$ induces a self-adjoint, positive definite operator with respect to T. Indeed, for $X, Y \in \mathcal{H}$, we have

$$T(A^\diamond(X), Y) = \frac{1}{2} \text{tr}(AXAY) = \frac{1}{2} \text{tr}(XAYA) = T(X, A^\diamond(Y)),$$

so $A^\diamond$ is self-adjoint, and we are left to showing $T(A^\diamond(X), X) = \frac{1}{2} \text{tr}(AXAX) > 0$. By (1.12) and the conjugate invariance of the trace, there is no loss of generality to assume that $A = \text{diag}(\beta, \beta^{-1})$ (note that $\det A = 1$). Let $X = \begin{pmatrix} \alpha & b \\ \bar{b} & \gamma \end{pmatrix}$, then

$$\text{tr}(AXAX) = \text{tr}\left(\begin{pmatrix} \beta\alpha & \beta b \\ \beta^{-1}\bar{b} & \beta^{-1}\gamma \end{pmatrix}^2 \right)$$

$$= \beta^2\alpha^2 + \beta^{-2}\gamma^2 + 2b\bar{b} > 0.$$

Hence $A^\diamond$ is positive definite. Therefore, the restrictions of ϕ

$$L_{\text{SL}(2,R)} \to L_{\text{SO}(2,1)} \quad \text{and,} \quad L_{\text{SL}(2,K)} \to L_{\text{SO}(3,1)},$$

respectively, are well-defined (i.e., they end up in $L_{\text{SO}(k,1)}$). By (2.7.6), they are homomorphisms of the K-loops.

$\phi(A) = \phi(B)$ for $B \in \mathrm{SL}(2, K) \cap \mathcal{H}$ entails $B = \pm A$ by (a). Exactly one of $A, -A$ is positive definite, hence the restrictions of ϕ are injective.

To deal with surjectivity, let $M = \begin{pmatrix} S & X \\ X^{\mathrm{T}} & T \end{pmatrix} \in L_{\mathrm{SO}(3,1)}$. The case of $L_{\mathrm{SO}(2,1)}$ runs exactly the same way, and is not worked out here. By (9.12.3), M is uniquely determined given $X = (x_1, x_2, x_3)^{\mathrm{T}} \in R^3$. To construct $A = \begin{pmatrix} \alpha & b \\ \bar{b} & \gamma \end{pmatrix} \in L_{\mathrm{SL}(2,K)}$ with $\phi(A^\circ) = M$, it therefore suffices to look at the action of A on the forth basis element:

$$\begin{pmatrix} \alpha & b \\ \bar{b} & \gamma \end{pmatrix} \begin{pmatrix} 1 & 0 \\ 0 & 1 \end{pmatrix} \begin{pmatrix} \alpha & b \\ \bar{b} & \gamma \end{pmatrix}^* = \begin{pmatrix} \alpha^2 + b\bar{b} & (\alpha + \gamma)b \\ (\alpha + \gamma)\bar{b} & \gamma^2 + b\bar{b} \end{pmatrix}.$$

Writing $b = b_1 + b_2 i$, this leads to the conditions

$$(x_1, x_2, x_3) = (\alpha + \gamma)\left(b_2, b_1, \frac{1}{2}(\alpha - \gamma)\right) \quad \text{and} \quad \alpha\gamma - b\bar{b} = 1.$$

Now M is in the image of our monomorphism (i.e., the restriction of ϕ) if and only if this system of equations has solutions $\alpha, \gamma \in R$, $b \in K$. To make this more handy, we rewrite it into

$$b_2 = \frac{x_1}{2v}, \quad b_1 = \frac{x_2}{2v}, \quad \frac{1}{2}(\alpha - \gamma) = \frac{x_3}{2v},$$

and

$$v^2 = 1 + b\bar{b} + \left(\frac{1}{2}(\alpha - \gamma)\right)^2,$$

where

$$v = \frac{1}{2}(\alpha + \gamma), \quad \text{thus} \quad \alpha = v + \frac{x_3}{2v}, \quad \gamma = v - \frac{x_3}{2v}.$$

The existence of α, γ, b is therefore equivalent with the existence of v, subject to

$$v^2 = 1 + \frac{x_1^2 + x_2^2 + x_3^2}{4v^2}, \quad \text{hence} \quad v^2 = \frac{1}{2}(1 + w),$$

where $w := \sqrt{1 + x_1^2 + x_2^2 + x_3^2} \in R$, $w > 0$.

If R is euclidean, there is a (unique) positive $v = \sqrt{(1/2)(1+w)} \in R$, and the existence of A follows. A will be positive definite, since $\det A = 1$ and the trace of A is $2v$.   ∎

Remark. By (1.26), the field $\mathbf{R}((t))$ is a pythagorean field, and using (1.25), the element

$$1 + \sqrt{1 + t^{-2}} = 1 + t^{-1}(1 + O(t^2)) = t^{-1}(1 + t + O(t^2))$$

is not a square. Hence in general pythagorean fields, $(1 + w)$ from the preceding proof is not necessarily a square.

We'll need conditions for the left inner mappings to be fixed point free:

(9.19) Let $\omega = \operatorname{diag}(U, V) \in \Omega_{U(p,q)} = U(p) \times U(q)$ and $A = \begin{pmatrix} S & X \\ X^* & T \end{pmatrix} \in L_{U(p,q)}$. Then we have $\omega A \omega^{-1} = A$ if and only if $UXV^* = X$. This implies $USU^{-1} = S$ and $VTV^{-1} = T$.

Proof. Noting that $U^* = U^{-1}$, $V^* = V^{-1}$, we find

$$\omega A \omega^{-1} = \begin{pmatrix} USU^{-1} & UXV^* \\ VX^*U^* & VTV^{-1} \end{pmatrix} \overset{!}{=} \begin{pmatrix} S & X \\ X^* & T \end{pmatrix}$$

implies all the given equations.

For the converse, assume that

$$UXV^* = X, \quad \text{hence} \quad X^* = (UXV^*)^* = VX^*U^*.$$

It remains to show that S and T are well behaved. (9.12.3) shows that $US^2U^{-1} = S^2$ and $VT^2V^{-1} = T^2$. Using the uniqueness of the square root (1.14), one can conclude that the same is true for S and T.   ∎

With this we can characterize those of our K-loops with fixed point free left inner mapping groups. Note that if R is euclidean, then the statement for $G = O(2, 1)$ follows from (9.18) and (9.3.2) already.

(9.20) Theorem. Let $G \in \{O(p,q), U(p,q)\}$ with $p \geq q$. Then the K-loop L_G has fixed point free left inner mappings if and only if $G \in \{O(1, 1), O(2, 1), U(1, 1)\}$.

Proof. The case $G = O(1,1)$ is trivial, since $L_{O(1,1)}$ is an abelian group by (9.17). In the case $G = U(1,1)$, the result is a consequence of (9.15) and (9.3.2).

The case $G = O(2,1)$ is also easy: In view of (9.16), the condition of (9.19) becomes $UX = X$ with $U \in SO(2)$, $X \in R^2$. But this is only possible when $U = I_2$.

For the converse, a proof for $G = U(2,1)$ and $G = O(p,q)$ with $p \geq 3$ suffices.

Let's do the case $G = O(p,q)$ first. We'll construct $A, B, C \in L_G$ such that $\delta_{A,B}(C) = C$, but $\delta_{A,B} \neq 1$. Then $\mathcal{D}(L_G)$ does not act fixed point free on L_G.

We begin with a simple matrix computation. Let

$$A' := \begin{pmatrix} \sqrt{2} & 0 & 1 \\ 0 & 1 & 0 \\ 1 & 0 & \sqrt{2} \end{pmatrix}, \quad B' := \begin{pmatrix} 1 & 0 & 0 \\ 0 & \sqrt{2} & 1 \\ 0 & 1 & \sqrt{2} \end{pmatrix} \in L_{O(2,1)},$$

then

$$A' \circ B' = \begin{pmatrix} \frac{5}{3} & \frac{\sqrt{2}}{3} & \sqrt{2} \\ \frac{\sqrt{2}}{3} & \frac{4}{3} & 1 \\ \sqrt{2} & 1 & 2 \end{pmatrix}$$

and

$$d_{A',B'} = \begin{pmatrix} \frac{2\sqrt{2}}{3} & \frac{1}{3} & 0 \\ -\frac{1}{3} & \frac{2\sqrt{2}}{3} & 0 \\ 0 & 0 & 1 \end{pmatrix}.$$

Indeed

$$A'B' = \begin{pmatrix} \sqrt{2} & 1 & \sqrt{2} \\ 0 & \sqrt{2} & 1 \\ 1 & \sqrt{2} & 2 \end{pmatrix} = (A' \circ B')d_{A',B'}.$$

Now let

$$A := \mathrm{diag}(I_{p-2}, A', I_{q-1}), \quad B := \mathrm{diag}(I_{p-2}, B', I_{q-1}) \in L_G,$$

then the previous calculation shows

$$d_{A,B} = \mathrm{diag}(I_{p-2}, d_{A',B'}, I_{q-1}), \quad \text{and clearly} \quad \hat{d}_{A,B} = \delta_{A,B} \neq 1.$$

Note that $d_{A,B}$ is of the form $\mathrm{diag}(U,V) \in \mathrm{SO}(p) \times \mathrm{SO}(q)$. Moreover, $V = I_q$. We are left with constructing C. To accomplish this, we employ (9.12.3). Put

$$X := \begin{pmatrix} 1 & 0 & \cdots & 0 \\ 0 & 0 & \cdots & 0 \\ \vdots & \vdots & \ddots & \vdots \\ 0 & 0 & \cdots & 0 \end{pmatrix} \in R^{p \times q} \quad \text{and let} \quad C := \begin{pmatrix} S & X \\ X^{\mathrm{T}} & T \end{pmatrix} \in L_G,$$

where S, T are uniquely determined given X. Recalling the hypotheses on p and q, it is easy to see that $UX = X = XV$. From (9.19) and (9.1.2) follows $\delta_{A,B}(C) = C$.

Finally, we'll do the case $G = \mathrm{U}(2,1)$. Recall that $L_{\mathrm{U}(p,q)} = L_{\mathrm{SU}(p,q)}$ by (9.14). The embedding $\Delta_2 : \mathrm{SL}(2,K) \to \mathrm{SL}(3,K)$ introduced just before (1.19) induces an embedding

$$\Delta : \begin{cases} \mathrm{SU}(1,1) \to \mathrm{SU}(2,1) \\ \qquad A \mapsto \mathrm{diag}(1,A) \end{cases}.$$

Using the isomorphism $\phi : \mathrm{SU}(1,1) \to \mathrm{SL}(2,R)$ of (9.15), from (2.7.6), (1.19) and (9.4) we can derive that

$$\omega := \mathrm{diag}(1,-1,-1)$$
$$= \Delta\phi^{-1}(-I_2) \in \Delta\phi^{-1}(\mathrm{SO}(2)) \subseteq \langle d_{A,B}; A, B \in L_G \rangle.$$

An easy calculation shows $\omega A' \omega^{-1} \neq A'$ and $\omega B' \omega^{-1} = B'$. Therefore $\widehat{\omega} \in \mathcal{D}(L_G)^{\#}$ has a fixed point $\neq I_3$. ∎

F. FIBRATIONS

Some of the K-loops constructed above posses a fibration. The key is hidden in the following definition. Let G be a subgroup of $\mathrm{GL}(n,K)$ such that $G = L_G\Omega_G$, i.e., L_G is a K-loop by (9.1.1). A

$\{I_n\} \neq L_H \neq L_G$, that for every $A \in L_G$ there exists $\omega \in \Omega_G$ with $\widehat{\omega}(A) \in L_H$, and for all $\omega \in \Omega_G$

$$\widehat{\omega}(L_H) \cap (L_H)^{\#} \neq \varnothing \implies \widehat{\omega}(L_H) \subseteq L_H.$$

With these conditions we get

(9.21) Theorem. Let G be a subgroup of $\mathrm{GL}(n, K)$ such that $G = L_G \Omega_G$, and let H be a fibration generating subgroup of G. Then $\mathcal{F} = \mathcal{F}_H := \{\widehat{\omega}(L_H); \ \omega \in \Omega_G\}$ is a $\widehat{\Omega}_G$-invariant (and thus $\mathcal{D}$-invariant) fibration of L_G. Moreover, all fibers, i.e., all elements of $\mathcal{F}$, are isomorphic to L_H.

Proof. By (2.7) L_H is a subloop of L_G. By (2.7.6) and (9.1.3) $\widehat{\omega} \in \mathrm{Aut}\, L_G$, thus all the fibers are isomorphic to L_H.

For $A \in L_G \setminus L_H$, choose $\omega \in \Omega_G$ with $\widehat{\omega}(A) \in L_H$. Then $A \in \widehat{\omega}^{-1}(L_H) \neq L_H$, therefore $|\mathcal{F}| \geq 2$, and $\cup_{F \in \mathcal{F}} F = L_G$.

Assume that $\widehat{\omega}_1(L_H) \cap \widehat{\omega}_2(L_H) \neq \{I_n\}$ for $\omega_1, \omega_2 \in \Omega$. Then for $\omega := \omega_2^{-1} \omega_1$ we have $\widehat{\omega}(L_H) \cap (L_H)^{\#} \neq \varnothing$, hence $\widehat{\omega}(L_H) \subseteq L_H$. We conclude that $\widehat{\omega}_1(L_H) \subseteq \widehat{\omega}_2(L_H)$, and by symmetry $\widehat{\omega}_1(L_H) = \widehat{\omega}_2(L_H)$. Therefore $\mathcal{F}$ is a fibration. ∎

We can now present some specific examples.

(9.22) Let $G \in \{\mathrm{SL}(2, R), \mathrm{SL}(2, K)\}$ and $H := \{\mathrm{diag}(\alpha, \alpha^{-1}); \ \alpha \in R, \alpha > 0\}$, then

(1) H is a fibration generating subgroup of G.

(2) $H = L_H$ and the fibration $\mathcal{F}_H$ consists of commutative subgroups of L_G, all isomorphic to H.

(3) For $A, B \in L_G$ with $A \neq I_2$ the following are equivalent

   (I) $A \circ B = B \circ A$, i.e., A, B commute in L_G;

   (II) $[A, B] \in \Omega_G$;

   (III) $[A, B] = I_2$, i.e., A, B commute in G;

   (IV) Let $F \in \mathcal{F}_H$ be the fiber containing A, i.e., $A \in F$, then $B \in F$.

Proof. (1) By (9.2) we have $G = L_G \Omega_G$, and $H = L_H \Omega_H$, since $L_H = H$, and $\Omega_H = \{I_2\}$. Clearly, $\{I_2\} \neq L_H \neq L_G$. By the spectral theorem (1.12) every $A \in L_G$ has a conjugate in $H = L_H$.

For the last requirement of the definition, assume that

$$A = \operatorname{diag}(\alpha, \alpha^{-1}) \in \hat{\omega}(H) \cap H^{\#} \quad \text{for some } \omega \in \Omega_G.$$

Now $\omega \in \operatorname{SU}(2, K)$ is of the form $\omega = \begin{pmatrix} a & b \\ -\bar{b} & \bar{a} \end{pmatrix}$. Having in mind that $\det \omega = a\bar{a} + b\bar{b} = 1$, we find for some $\beta > 0$.

$$\omega \operatorname{diag}(\beta, \beta^{-1})\omega^{-1} = \begin{pmatrix} a\bar{a}\beta + b\bar{b}\beta^{-1} & ab(-\beta + \beta^{-1}) \\ \bar{a}\bar{b}(-\beta + \beta^{-1}) & a\bar{a}\beta^{-1} + b\bar{b}\beta \end{pmatrix}$$

$$\overset{!}{=} \operatorname{diag}(\alpha, \alpha^{-1}),$$

Since $\beta = \beta^{-1}$ gives $\beta = 1$, which contradicts $A \neq I_2$, we can conclude that $ab = 0$. Therefore $\omega L_H \omega^{-1} \subseteq L_H$, and H is fibration generating.

(2) Clearly, H is a commutative subgroup of L_G. By (9.21) the other statements follow.

(3) "(I) $\Longrightarrow$ (II)" is in (2.7.5), and "(IV) $\Longrightarrow$ (I)" is in (2).

For the other two inclusions, there is no loss in generality to assume $A = \operatorname{diag}(\alpha, \alpha^{-1})$ with $\alpha > 1$, because of the spectral theorem (1.12). Let $B = \begin{pmatrix} \beta & b \\ \bar{b} & \gamma \end{pmatrix}$ with $\beta, \gamma \in R$, and $b \in K$. Of course, if $G = \operatorname{SL}(2, R)$, then $b = \bar{b} \in R$.

(II) $\Longrightarrow$ (III): By a direct computation we obtain that

$$[A, B] = \begin{pmatrix} \beta\gamma - \alpha^{-2}b\bar{b} & \gamma b(1 - \alpha^{-2}) \\ \beta\bar{b}(1 - \alpha^2) & \beta\gamma - \alpha^2 b\bar{b} \end{pmatrix} \in \Omega_G \subseteq \operatorname{SU}(2, K)$$

implies $\overline{\beta\gamma - \alpha^{-2}b\bar{b}} = \beta\gamma - \alpha^2 b\bar{b}$, hence $\alpha^4 = 1$, or $b = 0$. Since $\alpha > 1$, the first case is not possible. Thus $b = 0$, and $[A, B] = I_2$.

(III) $\Longrightarrow$ (IV): The fiber of A is H. If $[A, B] = I_2$, then the above calculation shows that $B \in H$. ∎

For the next example, we use the embedding

$$\Delta = \Delta_{n-1} : \operatorname{SL}(2, K) \to \operatorname{SL}(n, K); A \mapsto \operatorname{diag}(I_{n-2}, A)$$

introduced just before (1.19). In contrast with the preceding subsection E, we shall use $SO(p,1)$ and $SU(p,1)$ rather than the full pseudo-orthogonal and pseudo-unitary groups. Note that by (9.14) the corresponding loops are the same.

(9.23) Let $p = n - 1 > 1$ and $G \in \{SO(p,1), SU(p,1)\}$. We have

(1) $\Delta(SO(1,1))$ is a fibration generating subgroup of $SO(p,1)$. The fibers are commutative subgroups of $L_{O(p,1)}$.

(2) $\Delta(SU(1,1))$ is a fibration generating subgroup of $SU(p,1)$. The fibers are non-commutative subloops of $L_{U(p,1)}$, and are isomorphic to $L_{U(1,1)}$. In particular, they are proper K-loops.

The *Proof* is virtually the same for both cases, therefore only (2) will be worked out. Put $H := \Delta(SU(1,1))$.

Observe first that we really have an embedding

$$\Delta|_{SU(1,1)} : SU(1,1) \to SU(p,1) \quad \text{with} \quad \Delta(\Omega_{SU(1,1)}) \subseteq \Omega_{SU(p,1)}.$$

By (1.19) and (2.7.6) $\Delta|_{L_{U(1,1)}} : L_{U(1,1)} \to L_{U(p,1)}$ is an embedding of the K-loops. By (9.11) we have $H = L_H \Omega_H$. Clearly, $\{I_n\} \neq L_{O(1,1)} \subseteq L_{U(1,1)}$. The assumption $p > 1$ entails $L_H \neq L_G$.

For the two main conditions we take a closer look at H. Let $e_p = (0, \ldots, 0, 1)^T \in K^p$ be the pth unit vector. For $b \in K$ let $s = \sqrt{1 + b\bar{b}}$ (recall that R is pythagorean) and put

$$M(b) := \text{diag}\left(I_{n-2}, \begin{pmatrix} s & b \\ \bar{b} & s \end{pmatrix} \right)$$

$$= \begin{pmatrix} \text{diag}(1, \ldots, 1, s) & e_p b \\ \bar{b} e_p^T & s \end{pmatrix} \in L_H \quad \text{(see (9.12))}.$$

In fact $L_H = \{M(b); b \in K\}$. Note that according to (9.12.3) the matrix $M(b)$ as an element of L_G is uniquely determined by $e_p b$. For $\omega = \text{diag}(U, v) \in \Omega_G \subseteq U(p) \times U(1)$, we find

$$\hat{\omega}(A) = \omega \begin{pmatrix} S & X \\ X^* & t \end{pmatrix} \omega^{-1}$$

$$= \begin{pmatrix} USU^{-1} & UX\bar{v} \\ vX^*U^* & v\bar{v}t \end{pmatrix} \quad \text{for all} \quad A = \begin{pmatrix} S & X \\ X^* & t \end{pmatrix} \in L_G.$$

Notice that by (9.12) the chosen form of an element in L_G is most general. Recall that X determines A uniquely.

Let $A \in L_G$ be written as a block matrix as above. Then $X\|X\|^{-1}$ is a normalized vector in K^p and by (1.21) there exists $U \in \mathrm{SU}(p, K)$ with

$$U \frac{X}{\|X\|} = \mathbf{e}_p, \quad \text{hence} \quad UX \in \mathbf{e}_p K.$$

For $\omega := \mathrm{diag}(U, 1) \in \Omega_G$ we have $\widehat{\omega}(A) \in L_H$, because $\widehat{\omega}(A) \in L_G$ allows us to appeal to (9.12.3).

Finally assume for $\omega = \mathrm{diag}(U, v) \in \Omega_G$ and $a, b \in K^*$ that

$$\widehat{\omega}\big(M(a)\big) = M(b) \in \widehat{\omega}(L_H) \cap (L_H)^{\#}.$$

This implies $U\mathbf{e}_p a \bar{v} = \mathbf{e}_p b$, thus $U\mathbf{e}_p \in \mathbf{e}_p K$. Thus, $\widehat{\omega}\big(M(c)\big) \in L_H$ for all $c \in K$, and H is fibration generating.

The statements about the fibers are direct consequences of (9.21), together with (9.17), or (9.15) and (9.22), respectively. ∎

Remark. These examples are due to GABRIELI and KARZEL [36]. Note that this fact is not quite obvious, since their construction uses a geometric approach. In fact their construction is more general, because it can be applied with more ground fields. Producing example (2) they solved the problem whether fibrations in K-loops necessarily have commutative fibers to the negative.

10. Relativistic Velocity Addition

In this section, $\mathbf{R}$ will denote the field of real numbers, which is clearly an $(n+1)$-real ordered field for all $n \in \mathbf{N}$.[1] Our main concern will be the Lorentz group $O(3,1)$ of special relativity over $\mathbf{R}$. Because there is no difference in the exposition, we do the more general case $O(n,1)$, where $n \geq 2$ is the dimension of space.

We'll apply notation and results from §9, in particular from §9.E. To emphasize the physical meaning of some variables, we'll deviate slightly from notation introduced in §9. Let

$$\mathbf{R}_c^n := \{\mathbf{v} \in \mathbf{R}^n; \|\mathbf{v}\| < c\} \quad \text{for some } c > 0.$$

We'll later interpret the elements of $\mathbf{R}_c^n$ as admissible velocities with c the speed of light. There is no loss in generality to assume $c = 1$. We could also write $\mathbf{v} = \frac{\tilde{\mathbf{v}}}{c}$, $\tilde{\mathbf{v}} \in \mathbf{R}_c^n$, and substitute back in later.

A. LORENTZ BOOSTS

For $\mathbf{v} \in \mathbf{R}_1^n$ define

$$\gamma = \gamma(\mathbf{v}) := \frac{1}{\sqrt{1 - \mathbf{v}^T\mathbf{v}}} \quad \text{and} \quad \mathbf{B}(\mathbf{v}) := \begin{pmatrix} I_n + \frac{\gamma^2}{1+\gamma}\mathbf{v}\mathbf{v}^T & \gamma\mathbf{v} \\ \gamma\mathbf{v}^T & \gamma \end{pmatrix}.$$

Directly from the definition follows

$$\gamma^2 = 1 + \gamma^2\mathbf{v}^T\mathbf{v}. \tag{i}$$

$\mathbf{B}(\mathbf{v})$ is called the *Lorentz boost* of $\mathbf{v}$. Clearly, $\mathbf{B}(0) = I_{n+1}$. Recall the definition of $L_{O(n,1)}$ given at the beginning of §9. Notice that the elements of $L_{O(n,1)}$ are symmetric matrices. We have

(10.1) *The map* $\mathbf{B} : \mathbf{R}_1^n \to L_{O(n,1)}; \mathbf{v} \mapsto \mathbf{B}(\mathbf{v})$ *is a bijection.*

---

[1] Everything goes through with a euclidean $(n+1)$-real field, in particular with a real closed field.

Proof. By (1.13) the eigenvalues of $\mathbf{v}\mathbf{v}^{\mathrm{T}}$ are squares, therefore $I_n + \frac{\gamma^2}{1+\gamma}\mathbf{v}\mathbf{v}^{\mathrm{T}}$ is positive definite, and clearly symmetric. In order to apply (9.12), we compute using (i)

$$\left(I_n + \frac{\gamma^2}{1+\gamma}\mathbf{v}\mathbf{v}^{\mathrm{T}}\right)^2 = I_n + 2\frac{\gamma^2}{1+\gamma}\mathbf{v}\mathbf{v}^{\mathrm{T}} + \frac{\gamma^4}{(1+\gamma)^2}\mathbf{v}\mathbf{v}^{\mathrm{T}}\mathbf{v}\mathbf{v}^{\mathrm{T}}$$

$$= I_n + \mathbf{v}\mathbf{v}^{\mathrm{T}}\frac{\gamma^2}{(1+\gamma)^2}\left(2 + 2\gamma + \gamma^2\mathbf{v}^{\mathrm{T}}\mathbf{v}\right)$$

$$= I_n + \gamma^2\mathbf{v}\mathbf{v}^{\mathrm{T}}\frac{1}{(1+\gamma)^2}\left(1 + 2\gamma + \gamma^2\right)$$

$$= I_n + (\gamma\mathbf{v})(\gamma\mathbf{v})^{\mathrm{T}}.$$

Together with (i) and (9.12.3) this shows that $\mathbf{B}(\mathbf{v}) \in L_{O(n,1)}$ and that $\mathbf{B}$ is injective.

To see that $\mathbf{B}$ is surjective, we use notation from (9.12). For

$$A = \begin{pmatrix} S & X \\ X^{\mathrm{T}} & T \end{pmatrix} \in L_{O(n,1)}, \quad \text{put} \quad \mathbf{v} := \frac{1}{\sqrt{1+X^{\mathrm{T}}X}}X = \frac{1}{T}X,$$

then $\|\mathbf{v}\|^2 = \frac{X^{\mathrm{T}}X}{1+X^{\mathrm{T}}X} < 1$, thus $\mathbf{v} \in \mathbf{R}_1^n$, and we have

$$\gamma(\mathbf{v}) = \frac{1}{\sqrt{1-\mathbf{v}^{\mathrm{T}}\mathbf{v}}} = \frac{1}{\sqrt{1 - \frac{1}{1+X^{\mathrm{T}}X}X^{\mathrm{T}}X}} = \sqrt{1+X^{\mathrm{T}}X} = T$$

and $X = \gamma(\mathbf{v})\mathbf{v}$. Therefore the uniqueness statement from (9.12.3) implies $\mathbf{B}(\mathbf{v}) = A$. ∎

The map $\mathbf{B}$ can be used to carry the loop structure of $(L_{O(n,1)}, \circ)$ (see (9.1.1) and the explanation just before) over to $\mathbf{R}_1^n$. Thus $(\mathbf{R}_1^n, \bullet)$ becomes a K-loop, and $\mathbf{B}$ is an isomorphism. We'll now explicitly express the resulting operation "$\bullet$".

(10.2) Theorem. Let $\mathbf{v}_1, \mathbf{v}_2 \in \mathbf{R}_1^n$, and put $\gamma_k := \gamma(\mathbf{v}_k)$, $k = 1,2$. *Then*

$$\mathbf{v}_1 \bullet \mathbf{v}_2 = \frac{\mathbf{v}_1 + \mathbf{v}_2}{1 + \mathbf{v}_1^{\mathrm{T}}\mathbf{v}_2} + \frac{\gamma_1}{1+\gamma_1}\frac{\mathbf{v}_1\mathbf{v}_1^{\mathrm{T}}\mathbf{v}_2 - \mathbf{v}_2\mathbf{v}_1^{\mathrm{T}}\mathbf{v}_1}{1 + \mathbf{v}_1^{\mathrm{T}}\mathbf{v}_2},$$

i.e., $\mathbf{B}(\mathbf{v}_1) \circ \mathbf{B}(\mathbf{v}_2) = \mathbf{B}(\mathbf{v}_1 \bullet \mathbf{v}_2)$.

Moreover, $\mathbf{B}(\mathbf{v}_1)\mathbf{B}(\mathbf{v}_2) = \mathbf{B}(\mathbf{v}_1 \bullet \mathbf{v}_2)\,\mathrm{diag}(U,1)$ *with* $U \in SO(n)$. *Of course,* U *is uniquely determined by* $\mathbf{v}_1$ *and* $\mathbf{v}_2$.

Proof. By (2.7) (or (9.1)) and (10.1) we get

$$\mathbf{B}(\mathbf{v}_1)\mathbf{B}(\mathbf{v}_2) = \mathbf{B}(\mathbf{v}_1 \bullet \mathbf{v}_2)\omega$$

for a uniquely determined $\omega \in \Omega_{O(n,1)}$. By (9.16)

$$\omega = \mathrm{diag}(U,1), \quad U \in SO(n).$$

Therefore, the second assertion is proved, already. To get reasonable expressions, we write $\mathbf{v} = \mathbf{v}_1 \bullet \mathbf{v}_2$ and $\gamma = \gamma(\mathbf{v})$. Thus we obtain $\quad \mathbf{B}(\mathbf{v}_1)\mathbf{B}(\mathbf{v}_2) =$

$$= \begin{pmatrix} I_n + \frac{\gamma_1^2}{1+\gamma_1}\mathbf{v}_1\mathbf{v}_1^T & \gamma_1\mathbf{v}_1 \\ \gamma_1\mathbf{v}_1^T & \gamma_1 \end{pmatrix} \begin{pmatrix} I_n + \frac{\gamma_2^2}{1+\gamma_2}\mathbf{v}_2\mathbf{v}_2^T & \gamma_2\mathbf{v}_2 \\ \gamma_2\mathbf{v}_2^T & \gamma_2 \end{pmatrix}$$

$$= \begin{pmatrix} * & \gamma_2\mathbf{v}_2 + \frac{\gamma_1^2\gamma_2}{1+\gamma_1}\mathbf{v}_1\mathbf{v}_1^T\mathbf{v}_2 + \gamma_1\gamma_2\mathbf{v}_1 \\ * & \gamma_1\gamma_2(\mathbf{v}_1^T\mathbf{v}_2 + 1) \end{pmatrix}$$

$$\overset{!}{=} \mathbf{B}(\mathbf{v})\,\mathrm{diag}(U,1) = \begin{pmatrix} * & \gamma\mathbf{v} \\ * & \gamma \end{pmatrix}$$

where the "$*$" indicates a matrix entry, we don't really need to know. By the Schwarz inequality (1.8) we have

$$|\mathbf{v}_1^T\mathbf{v}_2| \le \|\mathbf{v}_1\|\|\mathbf{v}_2\| < 1 \quad \text{thus} \quad \gamma_1\gamma_2(\mathbf{v}_1^T\mathbf{v}_2 + 1) > 0.$$

This implies $\gamma = \gamma_1\gamma_2(1 + \mathbf{v}_1^T\mathbf{v}_2)$. Moreover, we derive from (i)

$$\gamma_1^{-1}(1 + \gamma_1) = 1 + \gamma_1^{-1} = 1 + \gamma_1 - \gamma_1\mathbf{v}_1^T\mathbf{v}_1$$

$$\Longrightarrow \gamma_1^{-1} = 1 - \frac{\gamma_1}{1+\gamma_1}\mathbf{v}_1^T\mathbf{v}_1.$$

Therefore, we find

$$\mathbf{v} = (1 + \mathbf{v}_1^T\mathbf{v}_2)^{-1}\left(\mathbf{v}_1 + \gamma_1^{-1}\mathbf{v}_2 + \frac{\gamma_1}{1+\gamma_1}\mathbf{v}_1\mathbf{v}_1^T\mathbf{v}_2\right)$$

$$= \frac{\mathbf{v}_1 + \mathbf{v}_2}{1 + \mathbf{v}_1^T\mathbf{v}_2} + \frac{\gamma_1}{1+\gamma_1}\frac{\mathbf{v}_1\mathbf{v}_1^T\mathbf{v}_2 - \mathbf{v}_2\mathbf{v}_1^T\mathbf{v}_1}{1 + \mathbf{v}_1^T\mathbf{v}_2}.$$

This is the result.   ∎

Remarks. 1. The formula for $v_1 \bullet v_2$ in the preceding theorem obviously implies $\mathbf{B}(\mathbf{v})^{-1} = \mathbf{B}(-\mathbf{v})$. This is also clear from the physics. Notice that inverses formed in the group $O(n, 1)$ and inverses formed in the loop $L_{O(n,1)}$ coincide by (9.1.1).

2. The first sources where the Lorentz boosts appear in the stated form seem to be [45] and [17]. BENZ [13] thus called them *Herglotz-Brill-matrices.*

3. In [12] BENZ generalizes the preceding theorem to arbitrary pre-Hilbert spaces.

B. SPECIAL RELATIVITY

In special relativity the n-dimensional world[2] is described by the vector space $\mathbf{R}^{n+1}$. It is an experimentally proved fact that the speed of light c is the same in every *inertial frame.* An element $(\mathbf{x}, ct) \in \mathbf{R}^{n+1}$ is referred to as an *event.* It is interpreted as a point $\mathbf{x}$ in space at a specified time t.

The space $\mathbf{R}^{n+1}$ carries the quadratic form

$$Q(\mathbf{x}, ct) := (\mathbf{x}^{\mathrm{T}}, ct) \operatorname{diag}(I_n, -1) \begin{pmatrix} \mathbf{x} \\ ct \end{pmatrix} = \mathbf{x}^{\mathrm{T}}\mathbf{x} - (ct)^2.$$

The isometry group of this space is $O(n, 1)$. The form Q induces the *Lorentz-Minkowski-distance d* of two events

$$d(a, b) := Q(b - a), \quad a, b \in \mathbf{R}^{n+1}.$$

The constancy of the speed of light implies that the Lorentz-Minkowski-distance of two event a, b measured in inertial frames S, S' is the same. A famous theorem of ALEXANDROV [13; A.6.1, p. 233] states, that a coordinate transformation is induced by

$$\begin{pmatrix} \mathbf{x}' \\ ct' \end{pmatrix} = A \begin{pmatrix} \mathbf{x} \\ ct \end{pmatrix} + h, \quad A \in O(n, 1), \ h \in \mathbf{R}^{n+1},$$

---

[2] In reality, of course, $n = 3$.

where $\mathbf{x}, t$ are measured in $\mathcal{S}$, and $\mathbf{x}', t'$ in $\mathcal{S}'$. By matching the origins of our coordinate systems we can assume that $h = 0$. This way we are left with the group $O(n, 1)$.

Assume now that the frame $\mathcal{S}$ moves with constant velocity $\tilde{\mathbf{v}}$ inside $\mathcal{S}'$, then $\tilde{\mathbf{v}} \in \mathbf{R}_c^n$. The origin of $\mathcal{S}$ has coordinates

$$\begin{pmatrix} \mathbf{0} \\ ct \end{pmatrix} \text{ in } \mathcal{S} \text{ and } \begin{pmatrix} \tilde{\mathbf{v}}t' \\ ct' \end{pmatrix} \text{ in } \mathcal{S}',$$

at time t, t', respectively. Since we do not allow reversion of time, we can assume $t, t' > 0$. Notice that $t = 0 \iff t' = 0$ by our choice of coordinates. Therefore, there exists $A \in O(n, 1)$ with

$$A \begin{pmatrix} \mathbf{0} \\ ct \end{pmatrix} = \begin{pmatrix} \tilde{\mathbf{v}}t' \\ ct' \end{pmatrix}.$$

Also

$$-c^2 t^2 = Q \begin{pmatrix} \mathbf{0} \\ ct \end{pmatrix} = Q \begin{pmatrix} \tilde{\mathbf{v}}t' \\ ct' \end{pmatrix} = (\tilde{\mathbf{v}}^T \tilde{\mathbf{v}} - c^2)t'^2,$$

hence

$$t' = \gamma t, \quad \text{where} \quad \gamma = \gamma(\mathbf{v}), \ \mathbf{v} = \frac{\tilde{\mathbf{v}}}{c}.$$

If we express A according to (9.12), we get

$$A = \begin{pmatrix} S & \gamma \mathbf{v} \\ Y & \gamma \end{pmatrix}.$$

Applying (9.11) and (10.1) there exist $\mathbf{B}(\mathbf{v}_1) \in L_{O(n,1)}$ and $\omega \in \Omega_{O(n,1)}$ with $A = \mathbf{B}(\mathbf{v}_1)\omega$. By (9.13)

$$\omega = \text{diag}(U, \pm 1) \in O(n) \times O(1).$$

Putting this together, we find

$$\begin{pmatrix} S & \gamma \mathbf{v} \\ Y & \gamma \end{pmatrix} = \mathbf{B}(\mathbf{v}_1) \begin{pmatrix} U & \\ & \pm 1 \end{pmatrix} = \begin{pmatrix} S & \pm \gamma(\mathbf{v}_1)\mathbf{v}_1 \\ Y & \pm \gamma(\mathbf{v}_1) \end{pmatrix}.$$

This implies $\mathbf{v} = \mathbf{v}_1$ and $\omega = \text{diag}(U, 1)$. Therefore ω acts only on the space coordinates. By choosing the respective coordinate axes of S and S' parallel (and with the same orientation), we can assume $U = I_n$. Then $A = \mathbf{B}(\mathbf{v})$ is a Lorentz boost. In this sense, relativistic velocities of frames are described by Lorentz boosts.

Now assume that there are three frames S, S', S'' with parallel respective axes of the same orientation, and matching origins. S moves inside S' with velocity $\mathbf{v}_2$, and S' moves inside S'' with velocity $\mathbf{v}_1$. From (10.2) we obtain, the velocity of S inside S'' is given by

$$\mathbf{v}_1 \bullet \mathbf{v}_2 = \frac{\mathbf{v}_1 + \mathbf{v}_2}{1 + \frac{\mathbf{v}_1^T \mathbf{v}_2}{c^2}} + \frac{\gamma_1}{1 + \gamma_1} \frac{1}{c^2} \frac{\mathbf{v}_1 \mathbf{v}_1^T \mathbf{v}_2 - \mathbf{v}_2 \mathbf{v}_1^T \mathbf{v}_1}{1 + \frac{\mathbf{v}_1^T \mathbf{v}_2}{c^2}}.$$

Indeed, the coordinate transformation S to S'' is

$$\mathbf{B}(\mathbf{v}_1)\mathbf{B}(\mathbf{v}_2) = \mathbf{B}(\mathbf{v}_1 \bullet \mathbf{v}_2) \operatorname{diag}(U, 1) \quad \text{for} \quad U \in \mathrm{SO}(n)$$

where we made use of (10.2).

Notice that the quasidirect product $L_{\mathrm{O}(n,1)} \times_Q (\mathrm{SO}(n) \times \mathrm{SO}(1))$ is the special orthochronous Lorentz group. This is a (normal) subgroup of index 4 inside $\mathrm{O}(n, 1)$, and the connected component of I_{n+1}. See also remark 1 after (9.17).

In the most interesting case $n = 3$, the matrix U induces a rotation of space, called the *Thomas rotation*.[3] See [11] or [116] for explicit formulas. Also in this case, the "addition formula" can be rewritten to

$$\mathbf{v}_1 \bullet \mathbf{v}_2 = \frac{\mathbf{v}_1 + \mathbf{v}_2}{1 + \frac{\mathbf{v}_1^T \mathbf{v}_2}{c^2}} + \frac{\gamma_1}{1 + \gamma_1} \frac{1}{c^2} \frac{\mathbf{v}_1 \times (\mathbf{v}_1 \times \mathbf{v}_2)}{1 + \frac{\mathbf{v}_1^T \mathbf{v}_2}{c^2}}.$$

Here "$\times$" denotes the cross product of $\mathbf{R}^3$.

Finally let us emphasize the upshot of this section: The set $\mathbf{R}_c^n$ of admissible velocities together with "$\bullet$", the relativistic addition of velocities, forms a K-loop. This comes about naturally as the left loop structure of a transversal (the set of boosts) of the coset space $\mathrm{O}(n, 1)/(\mathrm{O}(n) \times \mathrm{O}(1))$.

The credit for the discovery of the fact that $(\mathbf{R}_c^3, \bullet)$ is a K-loop is Ungar's (see [116] and the appendix).

---

[3] It is also called *Thomas precession* or *Wigner rotation*.

11. K-loops from the General Linear Groups over Rings

Let R be a commutative ring with 1, with unit group $E := E(R)$, and let I be an ideal of R. The set

$$GL_I(2, R) := \left\{ \begin{pmatrix} a & b \\ c & d \end{pmatrix}; ad - bc \in E, c \in I \right\}$$

clearly forms a group under usual matrix multiplication. If $I = \varepsilon R$ is a principal ideal, then we write briefly $GL_\varepsilon(2, R)$.

The Jacobson radical is denoted by $J(R)$.

(11.1) *If I is contained in $J(R)$, then for all $a, b, c, d \in R$*

$$\begin{pmatrix} a & b \\ c & d \end{pmatrix} \in GL_I(2, R) \iff a, d \in E \text{ and } c \in I.$$

Proof. If $a, d \in E$ and $c \in I$, then using (1.23) we find

$$ad - bc = \left(1 - bc(ad)^{-1}\right)ad \in (1 + I)E \subseteq E.$$

Conversely, $c \in I$ and $ad - bc \in E$ by definition. So there exists $\alpha \in E$ such that

$$1 = (ad - bc)\alpha = ad\alpha - bc\alpha \implies ad\alpha = 1 + bc\alpha \in E \quad \text{by (1.23)}.$$

Therefore $a, d \in E$. ∎

From now until the end of this section, assume that R has an involutory automorphism $x \mapsto \bar{x}$. Pick an element $\varepsilon \in J(R)$ with $\bar{\varepsilon} = \varepsilon$. Then

$$G := \left\{ \begin{pmatrix} a & b \\ \varepsilon b & \bar{a} \end{pmatrix}; a\bar{a} - \varepsilon b\bar{b} \in E \right\} \subseteq GL_\varepsilon(2, R)$$

is a group with a subgroup $\Omega := \{\mathrm{diag}(\alpha, \bar{\alpha}); \alpha \in E\}$. Notice that the map

$$\phi : E \to \Omega; \alpha \mapsto \mathrm{diag}(\alpha, \bar{\alpha})$$

is an isomorphism. Put

$$M(b) := \begin{pmatrix} 1 & b \\ \varepsilon \bar{b} & 1 \end{pmatrix} \quad \text{for } b \in R.$$

With $F := \{\alpha \in E;\ \bar{\alpha} = \alpha\}$, the units fixed under the involutory automorphism, we have

(11.2) Theorem. Let G, Ω be defined as above, and put $L := \{M(a);\ a \in R\}$. Then

(1) L is a transversal of the coset space G/Ω. In fact, $(L, \circ)$ is a K-loop, with

$$M(a) \circ M(b) = M\left(\frac{a+b}{1+\varepsilon \bar{a} b}\right), \quad M(a)^{\iota} = M(-a) \quad \text{and}$$

$$d_{M(a),M(b)} = \phi(1 + \varepsilon a \bar{b}) \quad \text{for all } a, b \in R,$$

where ι denotes the inverse of $M(a)$ with respect to "$\circ$".

(2) $\mathcal{D}(L) \cong (1 + \varepsilon R)/(F \cap (1 + \varepsilon R))$ and $E/F \cong \widehat{\Omega} \subseteq \operatorname{Aut} L$.

(3) $[M(a)] := \{X \in L;\ \delta_{M(a),X} = \mathbf{1}\} = \{M(b);\ b \in R,\ \varepsilon(a\bar{b} - \bar{a}b) = 0\}$ for all $a \in R$.

(4) L is a group if and only if $\varepsilon(a - \bar{a}) = 0$ for all $a \in R$.

(5) Let R be a domain and $\varepsilon \neq 0$. Then $\widehat{\Omega} \neq \{\mathbf{1}\}$ is a fixed point free transassociant of L, and $\mathcal{F} := \{[M(a)];\ a \in R\}$ is an Aut L-invariant fibration of L.

Proof. (1) For $\alpha, b \in R$ with $\alpha \in E$,

$$\begin{pmatrix} \alpha & b \\ \varepsilon \bar{b} & \bar{\alpha} \end{pmatrix} = \begin{pmatrix} 1 & b\bar{\alpha}^{-1} \\ \varepsilon \bar{b}\alpha^{-1} & 1 \end{pmatrix} \begin{pmatrix} \alpha & 0 \\ 0 & \bar{\alpha} \end{pmatrix} = M(b\bar{\alpha}^{-1}) \operatorname{diag}(\alpha, \bar{\alpha}).$$

By (11.1) the left hand matrix is a general element from G. Therefore, L is a transversal, since it is easy to see that the given decomposition is unique. This implies

$$M(a)M(b) = \begin{pmatrix} 1 & a \\ \varepsilon \bar{a} & 1 \end{pmatrix} \begin{pmatrix} 1 & b \\ \varepsilon \bar{b} & 1 \end{pmatrix} = \begin{pmatrix} 1 + \varepsilon a \bar{b} & a + b \\ \varepsilon(\bar{a} + \bar{b}) & 1 + \varepsilon \bar{a} b \end{pmatrix}$$

$$= M\left(\frac{a+b}{1+\varepsilon \bar{a} b}\right) \operatorname{diag}(1 + \varepsilon a \bar{b}, 1 + \varepsilon \bar{a} b),$$

hence $M(a) \circ M(b) = M\left(\dfrac{a+b}{1+\varepsilon \bar{a}b}\right)$, $M(a) \circ M(-a) = M(0) = I_2$,

and $d_{M(a),M(b)} = \mathrm{diag}(1 + \varepsilon a\bar{b}, 1 + \varepsilon \bar{a}b) = \phi(1 + \varepsilon a\bar{b})$.

To verify the Bol condition, we take $a, b \in R$, and obtain

$$
\begin{aligned}
M(a)M(b)M(a) &= \begin{pmatrix} 1 & a \\ \varepsilon \bar{a} & 1 \end{pmatrix} \begin{pmatrix} 1 & b \\ \varepsilon \bar{b} & 1 \end{pmatrix} \begin{pmatrix} 1 & a \\ \varepsilon \bar{a} & 1 \end{pmatrix} \\
&= \begin{pmatrix} 1 + \varepsilon(\bar{a}b + a\bar{b} + a\bar{a}) & 2a + b + \varepsilon a^2 \bar{b} \\ \varepsilon(2\bar{a} + \bar{b} + \varepsilon \bar{a}^2 b) & 1 + \varepsilon(\bar{a}b + a\bar{b} + a\bar{a}) \end{pmatrix} \\
&= M\left(\frac{2a + b + \varepsilon a^2 \bar{b}}{1 + \varepsilon(\bar{a}b + a\bar{b} + a\bar{a})}\right) (1 + \varepsilon(\bar{a}b + a\bar{b} + a\bar{a})) I_2 \, .
\end{aligned}
$$

This is an element of LFI_2, since $1 + \varepsilon(\bar{a}b + a\bar{b} + a\bar{a})$ is a unit of R, invariant under $x \mapsto \bar{x}$ (see (1.23)). Now $FI_2 \subseteq \Omega$ is clearly contained in the center of G, and thus in the core of Ω. From (3.12) we get that L is a Bol loop.

To see the automorphic inverse property we compute

$$
\begin{aligned}
M(a)^{\iota} \circ M(b)^{\iota} &= M(-a) \circ M(-b) \\
&= M\left(\frac{-a - b}{1 + \varepsilon \bar{a}b}\right) = M\left(\frac{a + b}{1 + \varepsilon \bar{a}b}\right)^{\iota} .
\end{aligned}
$$

(2) For $\alpha \in E$, $a \in R$, and $\omega := \mathrm{diag}(\alpha, \bar{\alpha})$ we compute

$$
\begin{aligned}
\hat{\omega}(M(a)) &= \begin{pmatrix} \alpha & 0 \\ 0 & \bar{\alpha} \end{pmatrix} \begin{pmatrix} 1 & a \\ \varepsilon \bar{a} & 1 \end{pmatrix} \begin{pmatrix} \alpha^{-1} & 0 \\ 0 & \bar{\alpha}^{-1} \end{pmatrix} \\
&= \begin{pmatrix} 1 & a\alpha\bar{\alpha}^{-1} \\ \varepsilon \bar{a}\bar{\alpha}\alpha^{-1} & 1 \end{pmatrix} = M(a\alpha\bar{\alpha}^{-1}).
\end{aligned}
$$

(i)

This immediately implies that $C_{\Omega}(L) = FI_2$. Since ϕ is an isomorphism, we can derive the second assertion from (2.8.6). Moreover, by (1) we have

$$
\{d_{M(a),M(b)}; \, a, b \in R\} = \phi(1 + \varepsilon R),
$$

and the first assertion follows, as well.

(3) $\delta_{M(a),M(b)} = 1 \iff 1 + \varepsilon a\bar{b} \in F \iff \varepsilon(a\bar{b} - \bar{a}b) = 0$.

(4) L is a group if and only if $\mathcal{D}(L) = \{1\}$. By (2) this is equivalent with $1 + \varepsilon R \subseteq F$, i.e., $1 + \varepsilon a = 1 + \varepsilon \bar{a}$ for all $a \in R$. Hence the result.

(5) $\varepsilon \neq 0$ implies by (4) that L is not a group, and also $\hat{\Omega} \neq \{1\}$. By (2.8.6) and (2.16) $\hat{\Omega}$ is a transassociant. By (i) $\hat{\omega}(M(a)) = M(a)$ if and only if $a\alpha\bar{\alpha}^{-1} = a$ that is $a(\alpha - \bar{\alpha}) = 0$. Since R is a domain, we can conclude that either $a = 0$, hence $M(a) = I_2$, or $\alpha = \bar{\alpha}$. In the latter case, $\omega = \alpha I_2 \in FI_2$ is in the center of G, hence $\hat{\omega} = 1$. It follows that $\hat{\Omega}$ is fixed point free.

Now let $a, b, c \in R$ with $\delta_{M(a),M(b)} = 1$, and $\delta_{M(a),M(c)} = 1$. Then

$$\varepsilon(a\bar{b} - \bar{a}b) = 0 = \varepsilon(a\bar{c} - \bar{a}c) \implies \bar{a}b\bar{c} = a\bar{b}\bar{c} = \bar{a}\bar{b}c$$
$$\implies b\bar{c} - \bar{b}c = 0.$$

Thus $\delta_{M(b),M(c)} = 1$, and (8.6) shows that $\mathcal{F}$ is an $\text{Aut } L$-invariant fibration.   ∎

We'll now give explicit examples and explore some of their properties. Recall that $K[[t]]$ is a local domain with $J(R) = tR$ if K is a field by (1.24).

(11.3) *Let K be a field, which has an involutory automorphism $\eta : x \mapsto \bar{x}$. Put $R := K[[t]]$, and extend η to R by $\eta(t) := t$. Construct the K-loop $(L, \circ)$ with $\varepsilon = t$, as in (11.2). We have*

(1) *$G := L \times_Q \hat{\Omega}$ is a Frobenius group with many involutions. Moreover, $\text{char } G = 2 \iff \text{char } K = 2$.*

(2) *If $\text{char } K \neq 2$, then L is uniquely 2-divisible and G is a specific group with $\text{char } G = 0$. Furthermore, $\mathcal{D}(L)$ contains no involutions. In particular, $\iota \notin \mathcal{D}(L)$.*

Proof. (1) By (11.2.5) and (7.4) G is a Frobenius group. Now let $a \in R$. We have

$$M(a) \circ M(a) = M\left(\frac{2a}{1 + \varepsilon a\bar{a}}\right) \overset{!}{=} I_2 \iff 2a = 0. \qquad \text{(ii)}$$

Therefore L is of exponent 2 if and only if $\text{char } K = 2$. In the case of $\text{char } K = 2$, (7.12) shows that L has many involutions.

Assume now char $K \neq 2$. There exists $\beta \in K$ with $\bar{\beta} \neq \beta$. Replacing β by $\beta - \bar{\beta}$ if necessary there is no loss in generality to require $\bar{\beta} = -\beta$. Let $\omega := \phi(\alpha)$, for $\alpha \in E$, then by (i) in the proof of the preceding theorem we have

$$\hat{\omega}\big(M(a)\big) = M(a\alpha\bar{\alpha}^{-1}) \overset{!}{=} M(-a) = M(a)^\iota \iff \bar{\alpha} = -\alpha.$$

Choosing $\alpha = \beta$ shows $\iota \in \hat{\Omega}$. If $\bar{\alpha} = -\alpha$, then

$$\overline{\alpha\beta^{-1}} = (-\alpha)(-\beta)^{-1} = \alpha\beta^{-1}.$$

Therefore $\alpha \in \beta F$, and $\hat{\omega} = \iota$ by (11.2.2). If $\hat{\omega}$ is any involution, then $(\alpha\bar{\alpha}^{-1})^2 = 1$, hence $\bar{\alpha} = -\alpha$, and ι is the only involution in $\hat{\Omega}$.

For every $a \in R$, the element $\big(M(a), \iota\big)$ of G is easily seen to be an involution. Therefore the involutions act transitively, and G is a Frobenius group with many involutions.[1]

(2) First, we show that $\mathcal{D}(L)$ contains no involutions, because this hooks up to the proof of (1). For $a \in R$, we have

$$\overline{1 + \varepsilon a} = 1 + \varepsilon\bar{a} = -(1 + \varepsilon a) \iff \varepsilon(a + \bar{a}) = -2.$$

This is impossible, since $\varepsilon(a + \bar{a})$ is in $J(R)$, while -2 is a unit. Therefore the assertion is a direct consequence of (11.2.2) and the description of ι in the last but one paragraph. Note that an involution in $\hat{\Omega}$ is necessarily equal to ι.

Next we show that L is uniquely 2-divisible. By (ii) L contains no element of order 2. Thus (6.8.2) shows that the square map is injective. To show surjectivity, we must find to every $b \in R$ an element $a \in R$ with

$$\frac{2a}{1 + \varepsilon a\bar{a}} = b \quad \text{(see (11.2.1)). Hence} \quad \frac{4a\bar{a}}{(1 + \varepsilon a\bar{a})^2} = b\bar{b}.$$

Rearranging terms, we find the quadratic equation for $a\bar{a}$

$$b\bar{b}\varepsilon^2(a\bar{a})^2 + (2b\bar{b}\varepsilon - 4)a\bar{a} + b\bar{b} = 0.$$

---

[1] Compare this with (7.14) and its proof.

Notice that this equation lives in $H[[t]]$, where H is the fixed field of η inside K. Of course, $H[[t]]$ is the fixed ring of η in R. The discriminant of this equation is

$$(2b\bar{b}\varepsilon - 4)^2 - 4(b\bar{b})^2\varepsilon^2 = 16(1 - b\bar{b}\varepsilon).$$

By (1.25), this is a square in $H[[t]]$, and

$$\sqrt{1 - b\bar{b}\varepsilon} = 1 - \frac{1}{2}b\bar{b}\varepsilon - \frac{1}{8}(b\bar{b}\varepsilon)^2 - \frac{1}{16}(b\bar{b}\varepsilon)^3 + O\left((b\bar{b}\varepsilon)^4\right).$$

Note that $b\bar{b}\varepsilon = 0 + O(t)$. Therefore one solution ξ (in the field of fractions of $H[[t]]$) of the quadratic equation in question is

$$\xi = \frac{2 - b\bar{b}\varepsilon - 2\sqrt{1 - b\bar{b}\varepsilon}}{b\bar{b}\varepsilon^2} = \frac{1}{4}b\bar{b} + \frac{1}{8}(b\bar{b})^2\varepsilon + b\bar{b}O\left((b\bar{b}\varepsilon)^2\right),$$

and $\xi \in H[[t]]$. (By the way, the other root is not in $H[[t]]$.) Then

$$a := \frac{1}{2}b(1 + \varepsilon\xi)$$

does the job, as a straightforward calculation shows.

Finally we show that G is a specific group with $\operatorname{char} G = 0$. In view of (7.19.2), all we have to show is that the order of $M(a)$ is infinite for all $a \in R \setminus \{0\}$.

For $n \in \mathbf{N}$, we'll show by induction that there exist polynomials $f_n, g_n \in \mathbf{Z}[x]$ such that

$$M(a)^{\underline{n}} = M\left(\frac{a(n + \xi f_n(\xi))}{1 + \xi g_n(\xi)}\right), \quad \text{with } \xi = \varepsilon a\bar{a}, \qquad \text{(iii)}$$

where $M(a)^{\underline{n}}$ denotes the n-th power of $M(a)$ in $(L, \circ)$. In fact, we'll calculate f_n and g_n explicitly soon. The case $n = 1$ is trivial ($f_1 = g_1 = 0$). Using the induction hypothesis and (11.2.1) we can compute

$$M(a)^{\underline{n+1}} = M(a) \circ M(a)^{\underline{n}} = M\left(\frac{a + \dfrac{a\big(n + \xi f_n(\xi)\big)}{1 + \xi g_n(\xi)}}{1 + \varepsilon\bar{a}\dfrac{a\big(n + \xi f_n(\xi)\big)}{1 + \xi g_n(\xi)}}\right)$$

$$= M\left(\frac{a\big(1 + \xi g_n(\xi) + n + \xi f_n(\xi)\big)}{1 + \xi\big(g_n(\xi) + n + \xi f_n(\xi)\big)}\right)$$

$$= M\left(\frac{a\big(n + 1 + \xi f_{n+1}(\xi)\big)}{1 + \xi g_{n+1}(\xi)}\right),$$

where

$$f_{n+1} = f_n + g_n, \quad g_{n+1} = x f_n + g_n + n, \quad \text{and} \quad f_1 = g_1 = 0.$$

This settles the case $\operatorname{char} K = 0$, because from (iii) the order of $M(a)$ in $(L, \circ)$ is infinity, unless $a = 0$.

For the rest of this proof, we can assume that $p := \operatorname{char} K$ is an odd prime. Next we prove that for all $n \in \mathbf{N}$ [2]

$$f_n = \sum_{k=1}^{\lfloor \frac{n-1}{2} \rfloor} \binom{n}{2k+1} x^{k-1} \quad \text{and} \quad g_n = \sum_{k=1}^{\lfloor \frac{n}{2} \rfloor} \binom{n}{2k} x^{k-1}. \qquad \text{(iv)}$$

Indeed, we compute by induction

$$f_n + g_n = \sum_{k=1}^{\lfloor \frac{n}{2} \rfloor} \left(\binom{n}{2k+1} + \binom{n}{2k}\right) x^{k-1}$$

$$= \sum_{k=1}^{\lfloor \frac{n}{2} \rfloor} \binom{n+1}{2k+1} x^{k-1} = f_{n+1}$$

---

[2] $\lfloor r \rfloor := k \in \mathbf{Z}$ is defined by $k \le r < k+1$ for all real numbers r.

and

$$xf_n + g_n + n = \sum_{k=2}^{\lfloor \frac{n-1}{2} \rfloor + 1} \binom{n}{2k-1} x^{k-1} + \sum_{k=1}^{\lfloor \frac{n}{2} \rfloor} \binom{n}{2k} x^{k-1} + n$$

$$= \sum_{k=1}^{\lfloor \frac{n-1}{2} \rfloor + 1} \binom{n}{2k-1} x^{k-1} + \sum_{k=1}^{\lfloor \frac{n}{2} \rfloor} \binom{n}{2k} x^{k-1}$$

$$= \sum_{k=1}^{\lfloor \frac{n+1}{2} \rfloor} \left(\binom{n}{2k-1} + \binom{n}{2k} \right) x^{k-1}$$

$$= \sum_{k=1}^{\lfloor \frac{n+1}{2} \rfloor} \binom{n+1}{2k} x^{k-1} = g_{n+1} \,.$$

Thus the sequences of polynomials f_n, g_n given in (iv) solve the recursion.

Now assume that $M(a)^{\underline{n}} = I_2$ for some $n \in \mathbf{N}$. Since ξ is transcendental over K, we must have $n \equiv 0 \mod p$, and $f_n = 0$. By (iv) this is only possible if all the binomial coefficients present in f_n are zero. But p is odd and divides n, so $n \geq 3$ and by (1.6) there is at least one non-zero coefficient of f_n. Therefore the order of $M(a)$ is infinite, unless $a = 0$, in this case as well. ∎

Remarks. 1. The construction in this section is due to KOLB and KREUZER [75]. They also suggested to using the ring of formal power series over the complex numbers, to obtain specific examples.

2. There are lots of fields with an involutory automorphism in any characteristic, e.g., finite fields of square order.

3. G. NAGY [91] uses the construction of (11.2) with some finite rings to obtain examples of K-loops of exponent 2 and of order 4^n for arbitrary $n \in \mathbf{N}$ which are generated by two elements. Hence the orders of such K-loops are not bounded. This solves the "restricted Burnside problem" for this class of loops.

12. Derivations

The method of derivation is due to DICKSON. It has been used in the past to construct nearfields and later quasifields from fields and skewfields.[1] KARZEL [52] axiomatized this method for groups replacing the skewfield. For an even more general setting see [121; II.1 p. 66]. Here we give a generalization which applies to constructing loops.

A. GENERAL THEORY

Let G be a group. A map $\phi : G \to \operatorname{Aut} G; \; a \mapsto \phi_a$ with $\phi_1 = 1$ is called a *weak derivation*. It is called a *derivation* if furthermore for all $a, b \in G$ there exists a unique $x \in G$ such that

$$x\phi_x(a) = b.$$

We have

(12.1) *Let G be a group with a weak derivation ϕ. If we let*

$$a \circ b := a\phi_a(b), \quad then$$

(1) $G^\phi := (G, \circ)$ *is a left loop. The identity elements of G and G^ϕ coincide. For all $a \in G$ the right inverse of a is given by $a' := \phi_a^{-1}(a^{-1})$, i.e., $a \circ a' = 1$.*

(2) ϕ *is a derivation if and only if G^ϕ is a loop.*

Proof. (1) The unique solution of the equation $a \circ x = b$ is given by $x = \phi_a^{-1}(a^{-1}b)$.

(2) is obvious. ∎

G^ϕ is called the *derived (left) loop*. We shall use this proposition without specific reference.

Remarks. 1. It is not necessary to assume the image of ϕ to be in $\operatorname{Aut} G$. The symmetric group of G would do. Derivations in the present sense have also been called *automorphic*.

---

[1] For the most general approach and some historic remarks see [112].

2. For weaker hypotheses in (12.1.2) if ϕ maps into a finite group, see [62; (2.6)].

To distinguish powers in G and G^ϕ, we denote powers in G^ϕ by $a^{\underline{k}}$, $k \in \mathbf{Z}$, e.g., $a^{\underline{3}} = a \circ (a \circ a)$. We list a bunch of straightforward properties.

(12.2) *Let G be a group and ϕ a weak derivation. We have*

(1) $\delta_{a,b} = \phi_{a \circ b}^{-1} \phi_a \phi_b$ *for $a, b \in G$. Therefore, $\mathcal{D}(G^\phi) \subseteq \operatorname{Aut} G$.*

(2) $\sigma \in \operatorname{Aut} G$ *is an automorphism of G^ϕ if and only if $\sigma \phi_a \sigma^{-1} = \phi_{\sigma(a)}$ for all $a \in G$.*

(3) G^ϕ *is left alternative if and only if $\phi_{a^2} = \phi_a^2$ for all $a \in G$.*

(4) G^ϕ *satisfies the left inverse property if and only if $\phi_{\phi_a^{-1}(a^{-1})} = \phi_a^{-1}$ for all $a \in G$. In this case $a^{\underline{-1}} = \phi_a^{-1}(a^{-1})$.*

(5) G^ϕ *is left power alternative if and only if $\phi_{a^{\underline{k}}} = \phi_a^k$ for all $a \in G$ and for all $k \in \mathbf{Z}$.*

(6) G^ϕ *is a Bol loop if and only if $\phi_{a \circ (b \circ a)} = \phi_a \phi_b \phi_a$ for all $a, b \in G$. In this case, ϕ is a derivation.*

(7) G^ϕ *is a group if and only if $\phi_{a \circ b} = \phi_a \phi_b$ for all $a, b \in G$. In this case, ϕ is a derivation.*

Proof. (1) For $a, b, x \in G$, we compute

$$a\phi_a(b)\phi_a\phi_b(x) = a\phi_a\big(b\phi_b(x)\big) = a \circ (b \circ x)$$
$$= (a \circ b) \circ \delta_{a,b}(x) = a\phi_a(b)\phi_{a \circ b}\delta_{a,b}(x).$$

Rearranging terms gives the result.

(2) For $a, b \in G$ we have $\sigma(a \circ b) = \sigma(a)\sigma\phi_a(b)$ and $\sigma(a) \circ \sigma(b) = \sigma(a)\phi_{\sigma(a)}\sigma(b)$. These are equal if and only if $\sigma\phi_a(b) = \phi_{\sigma(a)}\sigma(b)$. Hence the result.

(3) G^ϕ is left alternative if and only if $\delta_{a,a} = 1$ for all $a \in G$. Now (1) shows the assertion.

(4) For $a \in G$ put $a' := \phi_a^{-1}(a^{-1})$. Then G^ϕ satisfies the left inverse property if and only if $\delta_{a,a'} = 1$ for all $a \in G$, by (3.1.1). Now (12.1.1) and (1) show the result.

(5) Using (6.1.1) and (1), the result can be obtained easily.

(6) Let G^ϕ be a Bol loop, then by (6.4.1) and (1) we can compute

$$1 = \delta_{b,a}\delta_{a,boa} = \phi_{boa}^{-1}\phi_b\phi_a\phi_{ao(boa)}^{-1}\phi_a\phi_{boa}$$

$$\implies \phi_b\phi_a\phi_{ao(boa)}^{-1}\phi_a = 1.$$

Rearranging terms gives the assertion.

Conversely, for $a, b, c \in G$ we find

$$a \circ (b \circ (a \circ c)) = a\phi_a(b)\phi_a\phi_b(a)\phi_a\phi_b\phi_a(c)$$
$$= a \circ (b \circ a)\phi_{ao(boa)}(c) = (a \circ (b \circ a)) \circ c.$$

(3.10) shows that G^ϕ is a Bol loop, and thus ϕ is a derivation.

(7) is direct from (2.3) and (1).   ∎

Before we proceed with the theory, we give an example, which has been referred to in previous sections.

(12.3) Let $(G, +)$ be an abelian group, and put $\phi_0 = 1$ and $\phi_a = -1$ for all $a \in G^*$, i.e., $\phi_a(x) = -x$. Then

(1) G^ϕ is a left alternative left Kikkawa loop of exponent 2.

(2) If there exists an element $a \in G$ with $2a \neq 0$, i.e., if G is not of exponent 2, then G^ϕ is not a loop. In this case $\mathcal{D}(G^\phi) = \{\pm 1\}$.

(3) If no element of G has order 2, then $\mathcal{D}(G^\phi)$ is fixed point free.

Proof. (1) By (12.1.1) G^ϕ is a left loop. Clearly, $a \circ a = 0$ for all $a \in G$, hence (12.2.3) shows that G^ϕ is left alternative and of exponent 2. Since $a^{-1} = a$ for all $a \in G$, the left and automorphic inverse properties are trivial. (12.2.1) and (12.2.2) show A_ℓ, because -1 centralizes every automorphism of G, and $\phi_{-a} = \phi_a$ for all $a \in G$.

(2) Indeed, the equation $x \circ a = a$ has two solutions, namely 0 and $2a$. Moreover, by (12.2.1), $\delta_{2a,a} = \phi_a^{-1}\phi_{2a}\phi_a = -1$, because $2a \circ a = a$.

(3) is obvious.   ∎

Remark. In (3) one gets a Frobenius group $G^\phi \times_Q \mathcal{D}(G^\phi)$. However, this Frobenius group is isomorphic to the semidirect product of G with $\{\pm 1\}$. Indeed, the map

$$G \to G^\phi \times_Q \mathcal{D}(G^\phi); \; a \mapsto \begin{cases} (0,1) & \text{if } a = 0 \\ (a,-1) & \text{if } a \neq 0 \end{cases}$$

is a monomorphism, and $(0,-1)(a,-1)(0,-1)^{-1} = (-a,-1)$.

B. η-DERIVATIONS

For an epimorphism $\eta : G \to \overline{G}$, $\overline{G}$ a group, let

$$A_\eta := \{\alpha \in \operatorname{Aut} G; \; \eta\alpha = \eta\}.$$

Before we show how to use this to construct derivations, we record

(12.4) *Let G be a group and A a subset of $\operatorname{Aut} G$. Let N be the normal subgroup in G generated by the set $\{g^{-1}\alpha(g); \; g \in G, \alpha \in A\}$, and V an arbitrary normal subgroup in G. Then the following are equivalent*

   (I) $N \subseteq V$;

  (II)   *V is A-invariant and the action of A induced on G/V is trivial;*

 (III)   *For the canonical epimorphism $\eta : G \to G/V$ we have $A \subseteq A_\eta$.*

Proof. (I) $\Longrightarrow$ (II): If $\alpha \in A$, then for $g \in V$ we have $\alpha(g) \in gV = V$. Hence V is A-invariant. Furthermore, if $g \in G$, then $\alpha(g) \in gN \subseteq gV$. Thus $\alpha(gV) = gV$.

(II) $\Longrightarrow$ (III): For all $g \in G$, $\alpha \in A$ we have $\alpha(g)V = \alpha(gV) = gV$, thus $g^{-1}\alpha(g) \in V$. This implies $\eta(g^{-1}\alpha(g)) = 1$, and $\eta\alpha(g) = \eta(g)$. Hence $\alpha \in A_\eta$.

(III) $\Longrightarrow$ (I): For all $g \in G$, $\alpha \in A$:

$$\eta(g^{-1}\alpha(g)) = \eta(g^{-1})\eta\alpha(g) = \eta(g^{-1})\eta(g) = 1.$$

Therefore $g^{-1}\alpha(g) \in V$, and $N \subseteq V$. ∎

The following construction gives many derivations.

(12.5) Theorem. Let $G, \overline{G}$ be groups, and let $\eta : G \to \overline{G}$ be an epimorphism. For every map $\psi : \overline{G} \to A_\eta$ with $\psi_1 = 1$, we have

(1) $\phi := \psi\eta$ is a derivation, and for all $a \in G$ there exists a map $\mu_a : G \to \ker\eta$ such that $\phi_a(x) = x\mu_a(x)$.

(2) For all $a, b \in G$ and all $\alpha, \beta \in A_\eta$ we have $\phi_{\alpha(a)\beta(b)} = \phi_{ab} = \phi_{aob}$. Moreover, for all $k \in \mathbf{N} : \phi_{a^k} = \phi_{a^k}$.

(3) G^ϕ satisfies the left inverse property if and only if $\psi_{u^{-1}} = \psi_u^{-1}$ for all $u \in \overline{G}$ if and only if $\phi_{a^{-1}} = \phi_a^{-1}$ for all $a \in G$. In this case, $\phi_{a^{-1}} = \phi_{a^{-1}}$.

(4) G^ϕ is a Bol loop if and only if $\psi(uvu) = \psi_u\psi_v\psi_u$ for all $u, v \in \overline{G}$.

(5) G^ϕ is a group if and only if ψ is a homomorphism.

Proof. (1) The condition $\psi_1 = 1$ makes sure that $\phi_1 = 1$, hence G^ϕ is a left loop. For $a, b \in G$, consider the equation $x\phi_x(a) = b$. Applying η to both sides, gives $\eta(x) = \eta(ba^{-1})$, and then $x = b\phi_{ba^{-1}}(a^{-1})$. This is indeed a solution, since

$$\phi_x = \phi\big(b\phi_{ba^{-1}}(a^{-1})\big) = \psi\Big(\eta(b)\eta\big(\phi_{ba^{-1}}(a^{-1})\big)\Big)$$
$$= \psi\big(\eta(b)\eta(a^{-1})\big) = \psi\eta(ba^{-1}) = \phi_{ba^{-1}}.$$

Therefore $x = b\phi_{ba^{-1}}(a^{-1})$ is the unique solution of the equation in question.

For the last assertion, let $a, x \in L$. By (12.4) $x^{-1}\phi_a(x) \in \ker\eta$. This gives the result.

(2) $\phi_{\alpha(a)\beta(b)} = \psi\big(\eta\alpha(a)\eta\beta(b)\big) = \psi\big(\eta(a)\eta(b)\big) = \phi_{ab}$. Since $a \circ b = a\phi_a(b)$, the second equation now follows. This also implies the last statement.

(3) comes directly from (12.2.4).

(4) Assume $\psi(aba) = \psi_a\psi_b\psi_a$ for all $a, b \in \overline{G}$. Using (2) we can compute

$$\phi_{ao(boa)} = \phi_{aba} = \psi\big(\eta(a)\eta(b)\eta(a)\big)$$
$$= \psi\eta(a)\psi\eta(b)\psi\eta(a) = \phi_a\phi_b\phi_a.$$

By (12.2.6) G^ϕ is a Bol loop. The converse is a consequence of a similar calculation and again (12.2.6).

(5) From (12.2.7) and (2) the result can be deduced easily. ∎

Derivations constructed as in the theorem are called η-*derivations* with *factorization* $\phi = \psi\eta$. Note that the factorization is not unique. The map $\mu : G \to (\ker\eta)^G$ is called the *obstruction* of ϕ corresponding to η. It is unique given η, and it factors through η, more precisely: There exists a map $\nu : \overline{G} \to (\ker\eta)^G$ such that $\mu = \nu\eta$. This will be called the *factorization* of μ. If G is abelian, then the μ_a are homomorphisms. This gives a way to construct η-derivations on abelian groups.

(12.6) Let $G, \overline{G}$ be abelian groups, and let $\eta : G \to \overline{G}$ be an epimorphism with $U := \ker\eta$. Let $\nu : \overline{G} \to \mathrm{Hom}(G, U)$ be a map such that for all $a \in G$, $v \in \overline{G}$,

$$\nu_1(a) = 1, \quad \text{and} \quad U \subseteq \ker\nu_v.$$

Put $\mu := \nu\eta$, and $\phi_a := 1 + \mu_a$ for all $a \in G$ (i.e., $\phi_a(x) = x\mu_a(x)$ for all $x \in G$). Then ϕ is an η-derivation with corresponding obstruction μ. Moreover, we have

(1) Let $a \in G$. If $\mu_a(a^{-1}) = 1$, then $a^{\underline{-1}} = a^{-1}$.

(2) If $a^{\underline{-1}} = a^{-1}$, for all $a \in G$, then G^ϕ satisfies the automorphic inverse property if and only if $\mu_a = \mu_{a^{-1}}$ for all $a \in G$.

Proof. Let $a \in G$. Clearly, ϕ_a is an endomorphism of G. We first show that ϕ_a is bijective: For $b \in G$ put $x := b\mu_a(b^{-1})$. We compute

$$\begin{aligned}
\phi_a(x) &= x\mu_a(x) = b\mu_a(b^{-1})\mu_a\big(b\mu_a(b^{-1})\big) \\
&= b\mu_a(b^{-1})\nu_a\big(\eta(b)\eta\mu_a(b^{-1})\big) = b\mu_a(b^{-1})\nu_a\big(\eta(b)\big) \\
&= b\mu_a(b^{-1})\mu_a(b) = b.
\end{aligned}$$

Therefore ϕ_a is surjective.

Let $x \in G$ be in the kernel of ϕ_a, i.e., $1 = \phi_a(x) = x\mu_a(x)$. This implies

$$x = \mu_a(x)^{-1} \in U, \quad \text{and so} \quad \mu_a(x) = 1.$$

Thus ϕ_a has trivial kernel and is injective.

Now $\eta\phi_a = \eta(1 + \mu_a) = \eta + \eta\mu_a = \eta$. Therefore $\phi_a \in A_\eta$, and (12.5.1) shows the result.

(1) is direct from (12.2.4).

(2) $a^{-1} \circ b^{-1} = a^{-1} \phi_{a^{-1}}(b^{-1}) = a^{-1} \phi_a(b^{-1}) = (a \circ b)^{-1}.$ ∎

Remarks. 1. The group $\overline{G}$ does not play an essential role in the construction of η-derivations. It can always be replaced by $G/\ker \eta$.

2. The definition of η-derivations was inspired by ANDRÈ [3] and LÜNEBURG [86; p. 53f].

3. η-derivations have been used in [21] and [23] to construct Bol quasifields from fields.

Let G be an abelian group with a subgroup U and a map $\mu : G \mapsto$ Hom(G, U). The pair (U, μ) will be called a *derivation sprout* on G if μ factors through the canonical epimorphism $\eta : G \to G/U$, i.e., there exists a map $\nu : G/U \to$ Hom(G, U) such that $\mu = \nu\eta$. Moreover, we require that for all $a \in G$

$$\mu_1(a) = 1 \quad \text{and} \quad U \subseteq \ker \mu_a.$$

Notice that this is exactly what we looked at in the preceding theorem. Hence by $\phi_a := 1 + \mu_a$ for all $a \in G$, we obtain a derivation, the *derivation corresponding to* (U, μ).

We remark that ϕ determines μ by (12.5.1), while U is not unique in general, i.e., there might be distinct subgroups U, U' of G such that $(U, \mu), (U', \mu)$ are both derivation sprouts on G. The corresponding derivations are of course the same.

(12.7) Let ϕ be an η-derivation with factorization $\psi\eta$ on a group G, and let $\sigma \in \text{Aut}\, G$. We have

(1) If $\sigma \in A_\eta$, then $\sigma \in \text{Aut}\, G^\phi \iff \forall a \in G : \sigma\phi_a\sigma^{-1} = \phi_a$, i.e., σ centralizes $\phi(G)$.

(2) If A_η is abelian, then $A_\eta \subseteq \text{Aut}\, G^\phi$, and G^ϕ is an A_ℓ-loop.

Proof. (1) The condition $\eta\sigma = \eta$ implies $\phi_{\sigma(a)} = \phi_a$. Now (12.2.2) shows the assertion.

(2) comes directly from (1) and (12.2.2).   ∎

C. EXAMPLES

As a major application, we give a generalization of KREUZER's construction [79; (3.5)] of a left power alternative Kikkawa loop, which is not Bol. By (6.9) such a loop cannot be 2-divisible. In fact we can give some more examples to show independence of various axioms.

Let G, H be (additively written) abelian groups with the following properties. Assume G has a subgroup T of index 2, and T contains an element t of order 2. Let α be the endomorphism of G with kernel T and image $U := \{0, t\}$. These properties determine α uniquely. More specific

$$\alpha : \begin{cases} G \to G \\ x \mapsto \begin{cases} 0 & \text{if } x \in T; \\ t & \text{if } x \in G \setminus T. \end{cases} \end{cases}$$

Groups G with these properties are easy to find, in fact every finite abelian group with order divisible by 4 does the job. For a subset M of H with $0 \notin M$ put

$$\mu_{(a,b)} := \begin{cases} \alpha \times \mathbf{0} & \text{if } (a,b) \in T \times M; \\ \mathbf{0} \times \mathbf{0} & \text{if } (a,b) \in G \times H \setminus T \times M. \end{cases}$$

Recall that $\mathbf{0}$ denotes the zero homomorphism. Notice that $\mu_{(a,b)}$ can be viewed as an element of $\mathrm{Hom}\big(G \times H, T \times \{0\}\big)$.

(12.8) *With the notation just introduced, $(T \times \{0\}, \mu)$ is a derivation sprout on $G \times H$. Let ϕ be the corresponding derivation, then $L := (G \times H)^{\phi}$ is an A_ℓ-loop.*

(1) *L is a Kikkawa loop if and only if $M = -M$.*

(2) *L is left alternative if and only if $2b \notin M$ for all $b \in H$.*

(3) *L is a left power alternative Kikkawa loop if and only if $M = -M$ and for all $b \in H, n \in \mathbf{N}$,*

$$nb \in M \iff n \text{ is odd and } b \in M.$$

(4) *L is a Bol loop if and only if $2H + M = M$. These conditions imply that L is a K-loop.*

(5) L *is a group if and only if* $M = \varnothing$.

Proof. The construction guarantees that $(T \times \{0\}, \mu)$ is a derivation sprout. By (12.6) and (12.7.2) L is an A_ℓ-loop.

Before we go into the other proofs, we put down two useful properties of ϕ:

$$|1 + \alpha \times 0| = 2 \quad \text{hence for all } x, y \in L:$$

$$\phi_x = \phi_x^{-1} \quad \text{and} \quad \delta_{x,y} = \phi_{x+y}\phi_x\phi_y \,. \tag{i}$$

Indeed, the first statement follows from the obvious fact that $2\alpha = \alpha^2 = 0$, and implies the second. The third comes from (12.2.1), (12.5.2) and the second statement.

(1) Assume $M = -M$. For $(a, b) \in L$ we have

$$\mu_{(a,b)}(-a, -b) = (0,0), \quad \text{since } \alpha(-a) = 0 \text{ if } a \in T.$$

From (12.6.1) we get $(a, b)^{-1} = -(a, b)$. The hypotheses ensure us that

$$-(T \times M) = T \times M, \quad \text{so} \quad \mu_{-(a,b)} = \mu_{(a,b)}.$$

Thus (12.6.2) implies that L satisfies the automorphic inverse property. Moreover,

$$\phi_{(a,b)}^{-1} = \phi_{(a,b)} = 1 + \mu_{(a,b)} = 1 + \mu_{-(a,b)} = \phi_{-(a,b)}.$$

Thus by (12.5.3) L satisfies the left inverse property and is therefore a Kikkawa loop.

Conversely, if there exists $b \in M$ with $-b \notin M$, then

$$\phi_{(0,b)} = 1 + \alpha \times 0, \quad \text{while} \quad \phi_{(0,-b)} = 1.$$

Therefore (12.5.3) shows that L does not satisfy the left inverse property, and so is not a Kikkawa loop.

(2) For all $x = (a, b) \in L$ we find

$$\delta_{x,x} = \phi_{2x}\phi_x\phi_x = \phi_{(2a,2b)} = 1 \iff 2b \notin M,$$

because $2a \in T$. Since $\delta_{x,x} = 1$ for all $x \in L$ is equivalent with the left alternative property, we are done.

(3) Let L be a left power alternative Kikkawa loop. From (1) we get $M = -M$.

If $b \in H$, $n \in \mathbf{N}$ constitutes a counterexample to the displayed condition, assume that n is minimal. By (2), $n > 1$. Put $x := (0, b)$ and observe that for all $k \in \mathbf{N}$

$$\phi_{kx} \neq 1 \iff \phi_{kx} = 1 + \alpha \times 0 \iff kb \in M$$

A previous remark and (6.1.1) give

$$\phi_{(k+1)x}\phi_x\phi_{kx} = \delta_{x,x^k} = 1 \quad \text{for all } k \in \mathbf{N}.$$

Assume first $nb \in M$, then n is even, or $b \notin M$. If $b \notin M$, then

$$1 = \delta_{x,x^{n-1}} = \phi_{nx}\phi_x\phi_{(n-1)x} = \phi_{nx}\phi_{(n-1)x} \implies (n-1)b \in M.$$

Since n was minimal, we must have $b \in M$, a contradiction. Therefore $b \in M$. If n were even, then $n - 1$ would be odd and so $(n-1)b \in M$. But this implies $\delta_{x,x^{n-1}} \neq 1$, a contradiction as well.

Therefore we see that $nb \notin M$. This implies $b \in M$ and n odd, because we are looking at a counterexample. Since $n - 1$ is even and satisfies the displayed condition, we find $(n-1)x \notin T \times M$, and

$$\delta_{x,x^{n-1}} = \phi_{nx}\phi_x\phi_{(n-1)x} = \phi_x \neq 1,$$

the final contradiction. We conclude that there cannot exist a counterexample, thereby proving one direction.

For the converse, we see from (1) that L is a Kikkawa loop. Thus it remains to show that L is left power alternative. By (6.3), it suffices to prove for all $x = (a, b) \in L$, and $n \in \mathbf{N}$

$$\phi_{(n+1)x}\phi_x\phi_{nx} = \delta_{x,x^n} \overset{!}{=} 1.$$

Assume first that $\phi_{nx} \neq 1$, then $nx = (na, nb) \in T \times M$, and the assumptions imply that n is odd, and $b \in M$. Since T is of index 2, we also must have $a \in T$. Therefore $\phi_x = 1 + \alpha \times 0 = \phi_{nx}$. Finally, $n + 1$ is even, so $(n + 1)b \notin M$ and $\phi_{(n+1)x} = 1$. Consequently, $\delta_{x,x^n} = 1$ in this case.

Now for the case $\phi_{nx} = 1$: If $\phi_{(n+1)x} \neq 1$, then we can conclude as in the first case that $n + 1$ is odd, and $\phi_x = 1 + \alpha \times 0 = \phi_{(n+1)x}$, which entails the assertion.

So we are left with $\phi_{(n+1)x} = \phi_{nx} = 1$. If $a \notin T$, then $\phi_x = 1$ by definition. If $a \in T$, then also $na, (n + 1)a \in T$. Therefore $nb, (n + 1)b \notin M$. One of n and $n + 1$ is odd, and the assumptions imply $b \notin M$. Therefore $\phi_x = 1$, as well.

Summing up, we have seen that $\delta_{x,x^n} = 1$ for every $x \in L$.

(4) From (12.2.6), (12.5.2), and (i) we see that L is Bol if and only if

$$\phi_{2x+y} = \phi_{x+y+x} = \phi_x \phi_y \phi_x = \phi_y \quad \text{for all } x, y \in L.$$

Let $x = (a', b')$, $y = (a, b) \in G \times H$. Since $2a' \in T$, we have $2a' + a \in T \iff a \in T$. Therefore, L is Bol if and only if

$$2b' + b \in M \iff b \in M \quad \text{for all } b, b' \in L.$$

This implies $2H + M \subseteq M$. Since trivially $M \subseteq 2H + M$, we obtain $2H + M = M$ if L is Bol.

For the converse, assume $2H + M = M$ and let $b, b' \in H$. We get

$$b \in M \implies 2b' + b \in M$$

and
$$2b' + b \in M \implies b = 2(-b') + 2b' + b \in M.$$

As we have seen, this implies Bol.

If $b \in M$, then $-b = 2(-b) + b \in M$, and L is a Kikkawa loop, by (1). Hence the last assertion.

(5) If $M = \varnothing$, then $\phi_x = 1$ for all $x \in L$, and $L = (G \times H, +)$ is a group.

From (12.5.5) one easily sees that if L is a group, then ϕ is a homomorphism. If $M \neq \varnothing$, then $\ker \phi = G \times H \setminus T \times M$ would be a proper subgroup of $G \times H$. This is not the case, since for $a \in G \setminus T$ we have $2a \in T$.   ∎

There are examples for most of the possible combinations of axioms in the preceding theorem. All of the examples can be modified in many ways. We leave it to the reader to construct her favorite. The phrase "leads to" means using the construction of the preceding theorem with this H and M gives a loop with the specified properties.

(12.9) *We continue to use notation introduced just before the preceding lemma.*

(1) $H := \mathbf{Z}_3$, $M := \{2\}$, *leads to an* A_ℓ-*loop which is not a Kikkawa loop, and not left alternative.*

(2) $H := \mathbf{Z}_8$, $M := \{1\}$, *leads to a left alternative* A_ℓ-*loop which is not a Kikkawa loop.*

(3) $H := \mathbf{Z}_3$, $M := \{1,2\}$, *leads to a Kikkawa loop which is not left alternative.*

(4) $H := \mathbf{Z}$, $M := \{1,-1\}$, *leads to a left alternative Kikkawa loop which is not left power alternative.*

(5) $H := \mathbf{Z}_4 \times \mathbf{Z}_4$, $M := \{(1,2),(3,2)\}$, *leads to a left power alternative Kikkawa loop which is not Bol.*

(6) H *any abelian group,* $M := H \setminus 2H$, *leads to a K-loop, which is a group if and only if* $H = 2H$. *Examples with* $H \neq 2H$ *are any finite abelian group of even order,* $H := \mathbf{Z}$, *etc.*

Proof. (1), (2), and (3) are clear in view of (12.8).

(4) Conditions (1) and (2) of (12.8) are satisfied. Now $1 \in M$, $3 \notin M$ shows that the displayed condition of (12.8.3) is not true.

(5) Clearly, $-M = M$. Let $(b_1, b_2) \in H$, $n \in \mathbf{N}$, then

$$n(b_1, b_2) = (nb_1, nb_2) \in M \implies nb_1 \in \{1,3\} \implies n, b_1 \text{ odd},$$

and $nb_2 = 2$ with odd n implies $b_2 = 2$. Therefore $(b_1, b_2) \in M$. The converse is easy.

By (12.8.3) we are looking at a Kikkawa loop. We have $2(1,1) + (1,2) \notin M$, hence (12.8.4) shows that this loop is not Bol.

(6) Let $b, b' \in H$, then $2b' + b \in 2H \iff b \in 2H$, since $2H$ is a subgroup of H. This cannot be the case if $b \in M$, hence $2H + M \subseteq M$. The other inclusion is trivial, so (12.8.4) shows that we are looking at a K-loop. The rest is clear in view of (12.8.5) and well-known properties of abelian groups. ∎

Remarks. 1. Example (5) is KREUZER's [79; (3.5)]. He makes a slip in setting up his conditions. More specific, his (ii) is too weak. Indeed, this condition holds for our example (4), which is not left power alternative. However, all of KREUZER's examples, namely (5), $H = \mathbf{R}^*$, $M = \{-1\}$, and $H = \mathbf{Q}$, $M = \{2^{2k+1}; k \in \mathbf{Z}\}$, do satisfy the condition in (12.8.3). Therefore they qualify for (5).

2. $H := \mathbf{Z}_8$, $M := \{1, -1\}$, gives a finite example for (4).

In the construction of (12.6), if U has a complement in G, then we get a particularly simple setup.

(12.10) *Let G be an abelian group with a subgroup U which has a complement V, i.e., $G = U \oplus V$. Every map $\nu : V \to \mathrm{Hom}(V, U)$ with $\nu_0 := \mathbf{0}$ can be extended to a map $\mu : G \to \mathrm{Hom}(G, U)$ if we put for every $a \in G$*

$$\mu_a(u + v) := \nu_w(v), \quad \text{if } a \in w + U, u \in U, v \in V.$$

(1) *(U, μ) is a derivation sprout on G. Let ϕ be the corresponding derivation.*

(2) *If G is an elementary abelian 2-group, then G^ϕ is a Bol loop. G^ϕ is of exponent 2 if and only if $\nu_v(v) = 0$ for all $v \in V$.*

Proof. μ is well-defined, since $G = U \oplus V$.

(1) By construction μ factors through the canonical epimorphism $G \to G/U$. Clearly, $\mu_0 = \mathbf{0}$, and $U \subseteq \ker \mu_a$ for all $a \in G$.

(2) For all $a, b \in G$ we have

$$\phi_a \phi_b = (1 + \mu_a)(1 + \mu_b) = 1 + \mu_a + \mu_b + \mu_a \mu_b = 1 + \mu_a + \mu_b,$$

since $\mu_a \mu_b(G) \subseteq \mu_a(U) = 0$. This implies $\phi_a \phi_b = \phi_b \phi_a$ and $\phi_a^2 = 1$. Using this and (12.5.2) we can compute

$$\phi_{a \circ (b \circ a)} = \phi_{a+b+a} = \phi_b = \phi_a \phi_b \phi_a,$$

hence G^ϕ is Bol by (12.2.6). For the last assertion let $a = u + v$ with $u \in U$, $v \in V$. We find

$$a \circ a = a + \phi_a(a) = a + a + \mu_a(a) = \mu_a(a) = \nu_v(v).$$

So $a \circ a = 0$ if and only if $\nu_v(v) = 0$.   ∎

Remarks. 1. Part (1) has a converse: Namely, if (U, μ) is a derivation sprout, then a corresponding map ν can be found, which gives μ as in the theorem. The details will be left to the reader.

2. The map ν is very closely related to the map with the same name in (12.6).

Collected from the literature, here are some more

Examples. 1. [15] $G = \mathbf{Z}_2^3 = \mathbf{Z}_2 \oplus \mathbf{Z}_2^2$, and $\nu_w : \mathbf{Z}_2^2 \to \mathbf{Z}_2$; $(x, y) \mapsto w_1 w_2 y$, where $w = (w_1, w_2)$.

2. [100; Ex. 2] as above, but $\nu_w(x, y) = w_1(w_2 + 1)y$.

3. [77] $G = \mathbf{Z}_2^4 = \mathbf{Z}_2 \oplus \mathbf{Z}_2^3$, and $\nu_w : \mathbf{Z}_2^3 \to \mathbf{Z}_2$; $(x, y, z) \mapsto w_1 w_3 y + w_1 w_2 z$, where $w = (w_1, w_2, w_3)$.

The last two examples are of exponent 2, the first one is not.

The next example is a derivation, which is not an η-derivation. It gives a group if and only if R is associative.

4. [101] (dual to original) $\phi_a = \begin{pmatrix} 1 & & & & \\ 0 & 1 & & & \\ a_2 & 0 & 1 & & \\ 0 & a_2 & 0 & 1 & \\ a_4 & 0 & a_2 & 0 & 1 \end{pmatrix}$, $a \in R^5$,

where R is an alternative division ring. As mentioned earlier, this example has the property that all its loop isotopes are isomorphic, i.e., it is a G-loop.

In [103] ROBINSON gives a more general method to get Bol loops, which applies to certain subsets of a group. He proved that every Bol loop can be obtained in his manner. Unfortunately, ROBINSON's approach is not constructive, there is no indication how the group, the subset and the "generalized derivation" should be chosen.

Appendix: Some Remarks on the History

Quasigroups implicitly occured already when EULER conjectured the non-existence of certain mutually orthogonal Latin squares in 1782. However, EULER and his followers had a combinatorial point of view when studying Latin squares. It is very unlikely that they recognized an algebraic interpretation such as quasigroups.

The first instance, where a non-associative loop occurs is probably the multiplicative loop of the non-zero "Cayley numbers" or "octonions". GRAVES was the first to give a construction in 1843, only two months after HAMILTON's discovery of the quaternions. HAMILTON noted in 1844 that the multiplication was not associative. These results were published in 1848. In the meantime, CAYLEY, independently, found and published his construction. See [30; Ch. 9] for more historical details. Recall that the Cayley numbers $\mathbf{O}$ are *alternative*, i.e.,

$$a \cdot ab = aa \cdot b \quad \text{and} \quad ba \cdot a = b \cdot aa \quad \text{for all } a, b \in \mathbf{O},$$

which can be viewed as a weak form of the associative law. According to ZORN [126], ARTIN conjectured the alternative property for $\mathbf{O}$ and considered it a substitute for associativity. This took place in the 1920's.

DICKSON introduced in [28] what is now called "semifields" or "(non-associative) division rings", which also have a loop as the multiplicative structure, but this time without weak associativity. In that way he provided examples of loops, which are not Moufang. His paper seems to be the first instance where the axioms for a loop showed up rather explicitly. Actually, already the title of the paper hints at those axioms.

All these examples have in common that the loops are not the main concern. They merely show up as the multiplicative structure of an algebraic system with two binary operations. MOUFANG was the first to strip off the additive structure. In her famous paper [90] she derived from results of ZORN [126] that the alternative law in an algebra implies the *Moufang identities*

$$a(b \cdot ac) = (ab \cdot a)c \quad \text{and} \quad (ca \cdot b)a = c(a \cdot ba)$$
$$ab \cdot ca = (a \cdot bc)a \quad \text{and} \quad ab \cdot ca = a(bc \cdot a).$$

Then she considered loops satisfying these identities, now called *Moufang loops*,[1] and she proved

Moufang's Theorem. *Every Moufang loop is di-associative, i.e., any two elements generate a subgroup.* ∎

This seems to be the first structure theorem in loop theory. BOL [15] observed that the first Moufang identity implies the third and forth, and BRUCK [18] showed that they are all equivalent, see [20; VII, Lemma 3.1, p. 115] for a proof. Since the Moufang identities as displayed above, consist of two identities and their respective duals, the class of Moufang loops is *self-dual*, i.e., a loop is Moufang if and only if its dual is. As a consequence every identity proved for all Moufang loops entails its own dual.

It seems interesting to note that MOUFANG in her paper gave a definition which in modern terms is a "Bol loop with the right inverse property". In particular, she used the Bol identity.

BLASCHKE studied—in the framework of differential geometry—families of curves on surfaces in the 1920's. He inspired REIDE-MEISTER [97] and THOMSEN [111], and later BOL to study abstract "3-nets" or "webs" in the way HILBERT has developed his foundations of geometry. Results of this research were collected in [14].

In 1937 BOL discussed certain configurations in 3-nets. 3-nets can be constructed from groups, and THOMSEN [111] showed that a 3-net comes from a group if and only if the "Reidemeister config-uration" is valid. BOL [15] showed that certain weaker forms of the Reidemeister configuration lead to Moufang and to Bol loops. In the appendix to [15], he showed by example that there exist Bol loops which are not Moufang. BOL attributed this example to ZASSENHAUS.

A Bol loop is Moufang if and only if it is also right Bol. This statement is already in [15], and implicitly in [90]. The existence of proper Bol loops[2] therefore shows that the class of Bol loops is not self-dual.

A Moufang loop is a K-loop if and only if it is commutative. This

---

[1] She called them "quasigroups".

[2] "Proper" here means not Moufang.

follows easily from Moufang's theorem. In [15] there is an example, probably the first, for a commutative Moufang loop, which is not a group. BOL attributed this also to ZASSENHAUS.

The theory of Moufang loops, and in particular of commutative Moufang loops has been developed rather deeply by BRUCK, see [18] and his book [20]. The theory of finite Moufang loops culminates in the classification of all simple ones by LIEBECK [85]. Of course, this is built upon the classification of finite simple groups.

BRUCK himself gave a satisfying structure theory for commutative Moufang loops. See also the follow-up [110] by SMITH, which contains BRUCK's contributions and many interesting applications. We mention briefly one of them.

MANIN describes in [87; Ch. I] how the well-known construction of the "group law" on an elliptic curve can be generalized to cubic hypersurfaces. The outcome is a commutative Moufang loop.

As another application, Moufang loops play a role in the construction of the Monster sporadic finite simple group. This is described in [6].

The theory of K-loops is thus a continuation of this work, although not as far developed yet.

The notion of a K-loop occurs already in [20], unnamed though. [20; VII Thm. 5.2, p. 121] says that if the "core" [3] of a Moufang loop is a quasigroup, then this core is isotopic to a K-loop. BRUCK also gives a construction which applies to certain Moufang loops, see §6, in particular (6.14).

The systematic study of Bol loops has been started by ROBINSON in his thesis [99], which was published in [100]. His main results are displayed and partially generalized in §2 and §6. He also gave some results on K-loops, (6.12) and (6.13). GLAUBERMAN [40] studied certain K-loops, and obtained for his special case—among others— the theorem that these K-loops are A_ℓ-loops. The general result, which is crucial to the theory, has been proved independently in [34; 5.1], in [41; 3.12], and in [80], see also the remark after (6.6).

ROBINSON and GLAUBERMAN as well as KEPKA [57] give some

---

[3] This is not related to the core of a subgroup as defined in §1.

examples of (mostly finite) K-loops.

In the literature K-loops have also been called *Bruck loops*, but note that some authors require the additional property that the loop be uniquely 2-divisible. We'll explain later how these loops came by their present name.

In 1939 BAER [8] introduced the left loop structure on transversals of subgroups in groups. It was later proposed again by SABININ [105], [106]. SABININ also introduced the transassociant and the quasidirect product. In 1994 KREUZER and WEFELSCHEID [81] pinned down the properties a transversal must have to give a K-loop. These properties are used in §§9,11 to construct examples.

KIKKAWA [67] asked the question whether a reductive homogeneous space, which carries a "geodesic local multiplication" μ can be equipped with a global multiplication, coinciding locally with μ. He showed that a "homogeneous Lie loop" L is always a reductive space with a geodesic local multiplication. If L is a Kikkawa loop,[4] then μ is induced from the multiplication of the loop. One can show that Lie Kikkawa loops are in fact K-loops. Moreover, KIKKAWA develops a purely algebraic theory for Kikkawa loops, which are one subject of this book.

KIKKAWA's approach to questions in differential geometry was continued and enhanced by SABININ and his school. Results are collected in the recent book [107].

K-loops also play a role in the algebraization of absolute geometries. See [55] for an overview.

The original notion of a K-loop evolved from the study of neardomains F. In [53] KARZEL showed that the set

$$T_2(F) := \{x \mapsto a + bx;\ a, b \in F,\ b \neq 0\}$$

forms a sharply 2-transitive group in its natural action on F. And he showed the converse: every sharply 2-transitive group can be obtained this way. This is worked out in §7.D, see also [121; V §§1,2] for a different approach. KARZEL [54] gives a historically motivated

---

[4] This is our terminology, KIKKAWA called them "symmetric loops".

account on the connection between nearfields, neardomains and K-loops. Sharply 3-transitive groups can similarly be described by special neardomains, called "KT-fields". Sharply k-transitive groups for $k \geq 4$ have been long classified, see e.g. [93].

All presently known neardomains have associative addition, i.e., are *nearfields*. In order to settle the question whether proper neardomains exist, KERBY and WEFELSCHEID extracted the following conditions for the additive loop of a neardomain: $(F, +)$ satisfies the left and automorphic inverse properties, the precession maps are automorphisms (i.e., F is a Kikkawa loop), and $\delta_{a,b} = \delta_{a,b+a}$ for all $a, b \in F$. KERBY and WEFELSCHEID called a loop with these properties a *K-loop*.[5] However, they have used this notion only in talks in the 1970's and beginning 1980's, there doesn't seem to be a published source of that time. KIST [74; (1.8), p. 13] was probably the first who realized that K-loops in the above sense are Bol loops. KREUZER's theorem [80], see (6.7), shows that Bol loops with automorphic inverse property are K-loops. Only then it was clear that the older notion of a Bruck loop was the same as the notion of a K-loop.

In his thesis [35] GABRIEL introduced *specific groups*. GABRIEL's definition axiomatizes some properties of the subgroup of a sharply 2-transitive group, which is generated by the involutions. We generalize this notion, while keeping its flavor. In particular, every sharply 2-transitive group of characteristic $\neq 2$ is a specific group of the same characteristic, and so is every subgroup, which contains all the involutions.

The set of admissible velocities $\mathbf{R}_c^3 := \{ \mathbf{v} \in \mathbf{R}^3; \ |\mathbf{v}| < c \}$, where c is the speed of light, together with the relativistic velocity addition forms a K-loop. This has been proved by UNGAR (cf. [115, 116, 117]). Encouraged by UNGAR's discovery, many K-loops have been constructed in recent years (see, e.g., [48, 49, 56, 64, 75, 76, 77, 78, 79, 81]). In case of UNGAR's example $\mathbf{R}_c^3$, the precession maps are the "Thomas precessions" or "Thomas rotations" of special relativity, hence the name precession map.

The first published source, where the name "K-loop" has been men-

---

[5] The "K" is used in honor of KARZEL.

tioned is the introduction of UNGAR's paper [116]. Indeed, when WEFELSCHEID got the submitted paper in his hands in 1988, he immediately realized that the conditions UNGAR has found for the relativistic velocity addition match the properties KERBY and WE-FELSCHEID had extracted from the axioms of a neardomain to form the working definition "K-loop". Due to the lack of good examples KERBY and WEFELSCHEID did not publish their results, and the subject felt dormant until 1988. WEFELSCHEID immediately answered UNGAR, telling him in a letter that his discovery also was significant for the field of neardomains and sharply 2-transitive groups, and that he actually found a K-loop. In particular, WE-FELSCHEID pointed out the relevance of the unique solvability of the equations $ax = b$ and $ya = b$, which UNGAR then included into his paper. In later publications, UNGAR choose to call these structures gyrocommutative gyrogroups (see [119] for more). As mentioned earlier UNGAR's paper gave a decisive impetus to the subject.

References

[1]   A. A. ALBERT, *Quasigroups I*. Trans. Amer. Math. Soc. **54** (1943), 507–519.

[2]   A. A. ALBERT, *Quasigroups II*. Trans. Amer. Math. Soc. **55** (1944), 401–419.

[3]   J. ANDRÉ, *Über nicht-Desarguessche Ebenen mit transitiver Translationengruppe*. Math. Z. **60** (1954), 156–186.

[4]   R. ARTZY, *Relations between loop identities*. Proc. Amer. Math. Soc. **11** (1960), 847–851.

[5]   M. ASCHBACHER, *Finite Group Theory*. Cambridge Univ. Press, Cambridge 1986.

[6]   M. ASCHBACHER, *Sporadic Groups*. Cambridge Univ. Press, Cambridge 1994.

[7]   M. F. ATIYAH & I. G. MACDONALD, *Introduction to Commutative Algebra*. Addison-Wesley, Reading, Massachusetts 1969.

[8]   R. BAER, *Nets and groups*. Trans. Amer. Math. Soc. **46** (1939), 110–141.

[9]   E. BECKER, *Euklidische Körper und euklidische Hüllen von Körpern*. J. Reine Angew. Math. **268/269** (1974), 41–52.

[10]   V. D. BELOUSOV, *Foundations of the Theory of Quasigroups and Loops*. Izdat. Nauka, Moscow 1967, (Russian).

[11]   A. BEN-MENAHEM, *Wigner's rotation revisited*. Amer. J. Phys. **53** (1985), 62–66.

[12]   W. BENZ, *Lorentz-Minkowski distances in Hilbert spaces*. Geom. Dedicata **81** (2000), 219–230.

[13]   W. BENZ, *Geometrische Transformationen*. BI-Wissenschafts-Verlag, Mannheim-Wien-Zürich 1992.

[14]   W. BLASCHKE & G. BOL, *Geometrie der Gewebe*. Springer-Verlag, Berlin-Heidelberg-New York 1938.

[15] G. Bol, *Gewebe und Gruppen.* Math. Ann. **114** (1937), 414–431.

[16] A. Borovik & A. Nesin, *Groups of Finite Morley Rank.* Oxford Univ. Press, Oxford 1994.

[17] A. v. Brill, *Das Relativitätsprinzip.* Jahresber. Deutsch. Math.-Verein. **21** (1912), 60–87.

[18] R. H. Bruck, *Contributions to the theory of loops.* Trans. Amer. Math. Soc. **60** (1946), 245–354.

[19] R. H. Bruck & L. J. Paige, *Loops whose inner mappings are automorphisms.* Ann. of Math. (2) **63** (1956), 308–323.

[20] R. H. Bruck, *A Survey of Binary Systems*, 2nd ed. Springer-Verlag, Berlin-Heidelberg-New York 1966.

[21] R. P. Burn, *Bol quasi-fields and Pappus' theorem.* Math. Z. **105** (1968), 351–364.

[22] R. P. Burn, *Finite Bol loops.* Math. Proc. Cambridge Philos. Soc. **84** (1978), 377–385.

[23] A. Caggegi, *Nuovi quasicorpi di Bol.* Matematiche (Catania) **35** (1980), 241–247.

[24] O. Chein, H. O. Pflugfelder & J. D. H. Smith (eds.), *Quasigroups and Loops: Theory and Applications*, Heldermann Verlag, Berlin, 1990.

[25] A. C. Choudhury, *Quasi-groups and nonassociative systems* I. Bull. Calcutta Math. Soc. **40** (1948), 183–194.

[26] J. R. Clay, *Nearrings: Geneses and Applications.* Oxford Univ. Press, Oxford 1992.

[27] M. J. Collins, *Some infinite Frobenius groups.* J. Algebra **131** (1990), 161–165.

[28] L. E. Dickson, *Linear algebras in which division is always uniquely possible.* Trans. Amer. Math. Soc. (1906), 370–390.

[29] A. A. Drisko, *Loops with transitive automorphisms.* J. Algebra **184** (1996), 213–229.

[30] H.-D. EBBINGHAUS ET AL., *Zahlen*, 2nd ed. Springer-Verlag, Berlin-Heidelberg-New York 1988.

[31] D. EISENBUD, *Commutative Algebra with a View Toward Algebraic Geometry.* Springer-Verlag, Berlin-Heidelberg-New York 1995.

[32] N. J. FINE, *Binomial coefficients modulo a prime.* Amer. Math. Monthly **54** (1947), 589–592.

[33] L. FUCHS, *Infinite Abelian Groups.* Academic Press, New York-London 1970.

[34] M. FUNK & P. T. NAGY, *On collineation groups generated by Bol reflections.* J. Geom. **48** (1993), 63–78.

[35] C. M. GABRIEL, *Verallgemeinerungen scharf zweifach transitiver Permutationsgruppen und das Burnside-Problem.* Ph. D. thesis, Univ. Hamburg, 1997.

[36] E. GABRIELI & H. KARZEL, *Point-reflection geometries, geometric K-loops and unitary geometries.* Resultate Math. **32** (1997), 66–72.

[37] E. GABRIELI & H. KARZEL, *Reflection geometries over loops.* Resultate Math. **32** (1997), 61–65.

[38] E. GABRIELI & H. KARZEL, *The reflection structures of generalized co-Minkowski spaces leading to K-loops.* Resultate Math. **32** (1997), 73–79.

[39] The GAP Group, Aachen, St Andrews, GAP—*Groups, Algorithms, and Programming*, 1999, http://www-gap.dcs.st-and.ac.uk/~gap.

[40] G. GLAUBERMAN, *On loops of odd order.* J. Algebra **1** (1964), 374–396.

[41] E. G. GOODAIRE & D. A. ROBINSON, *Semi-direct products and Bol loop.* Demonstratio Math. **27** (1994), 573–588.

[42] L. C. GROVE, *Groups and Characters.* Wiley-Interscience, New York 1997.

[43] W. HEIN, *Einführung in die Struktur- und Darstellungs-theorie der klassischen Gruppen.* Springer-Verlag, Berlin-Heidelberg-New York 1990.

[44] S. HELGASON, *Differential Geometry and Symmetric Spaces.* Academic Press, New York-London 1962.

[45] G. HERGLOTZ, *Über die Mechanik des deformierbaren Körpers vom Standpunkt der Relativitätstheorie.* Ann. Physik **36** (1911), 493–533, in "Gesammelte Schriften", Vandenhoeck & Ruprecht, Göttingen 1979.

[46] K.-H. HOFMANN & K. STRAMBACH, *Topological and analytic loops,* In Chein et al. [24], pp. 205–262.

[47] B. HUPPERT, *Endliche Gruppen* I. Springer-Verlag, Berlin-Heidelberg-New York 1967.

[48] B. IM, *K-loops in the Minkowski world over an ordered field.* Resultate Math. **25** (1994), 60–63.

[49] B. IM, *K-loops and quasidirect products in 2-dimensional linear groups over a pythagorean field.* Resultate Math. **28** (1995), 67–74.

[50] B. IM & H.-J. KO, *Web loops and webs with reflections.* J. Geom. **61** (1998), 62–73.

[51] K. W. JOHNSON & B. L. SHARMA, *A variety of loops.* Ann. Soc. Sci. Bruxelles Sér. I **92** (1978), 25–41.

[52] H. KARZEL, *Unendliche Dicksonsche Fastkörper.* Arch. Math. (Basel) **16** (1965), 247–256.

[53] H. KARZEL, *Zusammenhänge zwischen Fastbereichen, scharf zweifach transitiven Permutationsgruppen und 2-Strukturen mit Rechtecksaxiom.* Abh. Math. Sem. Univ. Hamburg **32** (1968), 191–206.

[54] H. KARZEL, *From nearrings and nearfields to K-loops,* In Saad and Thomsen [104], pp. 1–20.

[55] H. KARZEL, *Recent developments on absolute geometries and algebraization by K-loops.* Discrete Math. **208/209** (1999), 387–409.

[56] H. KARZEL & H. WEFELSCHEID, *Groups with an involutory antiautomorphism and K-loops; Application to space-time-world and hyperbolic geometry* I. Resultate Math. **23** (1993), 338–354.

[57] T. KEPKA, *A construction of Bruck loops.* Comment. Math. Univ. Carolin. **25** (1984), 591–595.

[58] W. KERBY, *Sharply 3-transitive groups of characteristic $\equiv 1$ mod 3.* J. Algebra **32** (1974), 240–245.

[59] W. KERBY & H. WEFELSCHEID, *Bemerkungen über Fastbereiche und scharf zweifach transitive Gruppen.* Abh. Math. Sem. Univ. Hamburg **37** (1972), 20–29.

[60] W. KERBY & H. WEFELSCHEID, *Über eine scharf 3-fach transitiven Gruppen zugeordnete algebraische Struktur.* Abh. Math. Sem. Univ. Hamburg **37** (1972), 225–235.

[61] E. I. KHUKHRO, *Nilpotent Groups and their Automorphisms.* Walter de Gruyter, Berlin-New York 1993.

[62] H. KIECHLE, *Lokal endliche Quasikörper.* Ph. D. thesis, Techn. Univ. München, 1990.

[63] H. KIECHLE, *Der Kern einer automorphen Ableitung und eine Anwendung auf normale Teilkörper verallgemeinerter André-Systeme.* Arch. Math. (Basel) **58** (1992), 514–520.

[64] H. KIECHLE, *K-loops from classical groups over ordered fields.* J. Geom. **61** (1998), 105–127.

[65] H. KIECHLE & A. KONRAD, *The structure group of certain K-loops,* In Saad and Thomsen [104], pp. 287–294.

[66] M. KIKKAWA, *On some quasigroups of algebraic models of symmetric spaces* II. Mem. Fac. Sci. Shimane Univ. **7** (1974), 29–35.

[67] M. KIKKAWA, *Geometry of homogeneous Lie loops.* Hiroshima Math. J. **5** (1975), 141–179.

[68] M. KIKKAWA, *On some quasigroups of algebraic models of symmetric spaces* III. Mem. Fac. Sci. Shimane Univ. **9** (1975), 7–12.

[69] M. K. KINYON, *personal communication*, June 2001.

[70] M. K. KINYON, *private communication*, Email February 2001.

[71] M. K. KINYON, *private communication*, Email September 1999.

[72] M. K. KINYON, *Global left loop structure on spheres.* Comment. Math. Univ. Carolin. **41** (2000), 325–346.

[73] M. K. KINYON & O. JONES, *Loops and semidirect products.* Comm. Algebra **28** (2000), 4137–4164.

[74] G. KIST, *Theorie der verallgemeinerten kinematischen Räume.* Beiträge zur Geometrie und Algebra **14** (1986), TUM-M8611, Habilitationsschrift, Techn. Univ. München.

[75] E. KOLB & A. KREUZER, *Geometry of kinematic K-loops.* Abh. Math. Sem. Univ. Hamburg **65** (1995), 189–197.

[76] A. KONRAD, *Hyperbolische Loops über Oktaven und K-Loops.* Resultate Math. **25** (1994), 331–338.

[77] A. KREUZER, *Beispiele endlicher und unendlicher K-Loops.* Resultate Math. **23** (1993), 355–362.

[78] A. KREUZER, *K-loops and Bruck loops on* $\mathbf{R} \times \mathbf{R}$. J. Geom. **47** (1993), 86–93.

[79] A. KREUZER, *Construction of finite loops of even order.* Proc. of the Conference on Nearrings and Nearfields (Fredericton, NB, Canada, 18. – 24. July 1993) (Y. Fong & al., eds.), Kluwer Acad. Press, 1995, pp. 169–179.

[80] A. KREUZER, *Inner mappings of Bol loops.* Math. Proc. Cambridge Philos. Soc. **123** (1998), 53–57.

[81] A. KREUZER & H. WEFELSCHEID, *On K-loops of finite order.* Resultate Math. **25** (1994), 79–102.

[82] K. KUNEN, *Moufang quasigroups.* J. Algebra **183** (1996), 231–234.

[83] R. LAL, *Transversals in groups.* J. Algebra **181** (1996), 70–81.

[84] S. LANG, *Algebra*, 2nd ed. Addison-Wesley, Reading, Massachusetts 1984.

[85] M. W. LIEBECK, *The classification of finite simple Moufang loops*. Math. Proc. Cambridge Philos. Soc. **102** (1987), 33–47.

[86] H. LÜNEBURG, *Translation Planes*. Springer-Verlag, Berlin-Heidelberg-New York 1980.

[87] Y. MANIN, *Cubic Forms*. North-Holland, Amsterdam-New York 1974.

[88] P. O. MIHEEV & L. V. SABININ, *Quasigroups and differential geometry*, In Chein et al. [24], pp. 357–430.

[89] D. MORNHINWEG, D. B. SHAPIRO & K. G. VALENTE, *The principal axis theorem over arbitrary fields*. Amer. Math. Monthly **100** (1993), 749–754.

[90] R. MOUFANG, *Zur Struktur von Alternativkörpern*. Math. Ann. **110** (1935), 416–430.

[91] G. P. NAGY, *Burnside problems for Moufang and Bol loops of small exponent*. Acta Sci. Math. (Szeged) **67** (2001), 471–480, to appear.

[92] L. J. PAIGE, *A class of simple Moufang loops*. Proc. Amer. Math. Soc. **7** (1956), 471–482.

[93] D. S. PASSMAN, *Permutation Groups*. W. A. Benjamin, New York-Amsterdam 1998.

[94] H. O. PFLUGFELDER, *Quasigroups and Loops: Introduction*. Heldermann-Verlag, Berlin 1990.

[95] G. PICKERT, *Projektive Ebenen*, 2nd ed. Springer-Verlag, Berlin-Heidelberg-New York 1975.

[96] V. V. PRASOLOV, *Problems and Theorems in Linear Algebra*, Translations of mathematical monographs, vol. 134. Amer. Math. Soc. 1994.

[97] K. REIDEMEISTER, *Gewebe und Gruppen*. Math. Z. **29** (1929), 427–435.

[98] J. R. RETHERFORD, *Hilbert Space: Compact Operators and the Trace Theorem.* Cambridge Univ. Press, Cambridge 1993.

[99] D. A. ROBINSON, *Bol loops.* Ph. D. thesis, Univ. Wisconsin, 1964.

[100] D. A. ROBINSON, *Bol loops.* Trans. Amer. Math. Soc. **123** (1966), 341–354.

[101] D. A. ROBINSON, *A Bol loop isomorphic to all loop isotopes.* Proc. Amer. Math. Soc. **19** (1968), 671–672.

[102] D. A. ROBINSON, *Bol quasigroups.* Publ. Math. Debrecen **19** (1972), 151–153.

[103] D. A. ROBINSON, *A special embedding of Bol loops in groups.* Acta Math. Hungar. **30** (1977), 95–103.

[104] G. SAAD & M. THOMSEN (eds.), *Nearrings, Nearfields and K-loops*, Proc. of the Conference on Nearrings and Nearfields (Hamburg, 30. 7. – 6. 8. 1995), Kluwer Acad. Press, 1997.

[105] L. V. SABININ, *Loop geometries.* Math. Notes **12** (1972), 799–805.

[106] L. V. SABININ, *On the equivalence of categories of loops and homogeneous spaces.* Soviet Math. Dokl. **13** (1972), 970–974.

[107] L. V. SABININ, *Smooth Quasigroups and Loops.* Kluwer Academic Publishers, Doldrecht-Boston-London 1999.

[108] L. V. SABININ, L. L. SABININA & L. V. SBITNEVA, *On the notion of gyrogroup.* Aequationes Math. **56** (1998), 11–17.

[109] B. L. SHARMA, *Left loops which satisfy the left Bol identity.* Proc. Amer. Math. Soc. **61** (1976), 189–195.

[110] J. D. H. SMITH, *Commutative Moufang loops: The first 50 years.* Algebras Groups Geom. **3** (1985), 209–234.

[111] G. THOMSEN, *Schnittpunktssätze in ebenen Geweben.* Abh. Math. Sem. Univ. Hamburg **7** (1930), 99–106.

[112] J. TIMM, *Zur Konstruktion von Fastringen* I. Abh. Math. Sem. Univ. Hamburg **35** (1970), 57–74.

[113] J. TITS, *Sur les groupes doublement transitifs continus*. Comment. Math. Helv. **26** (1952), 203–224.

[114] J. TITS, *Sur les groupes doublement transitifs continus: correction et compléments*. Comment. Math. Helv. **30** (1955), 234–240.

[115] A. A. UNGAR, *Thomas rotation and the parametrization of the Lorentz transformation group*. Found. Phys. Lett. **1** (1988), 57–89.

[116] A. A. UNGAR, *The relativistic noncommutative nonassociative group of velocities and the Thomas rotation*. Resultate Math. **16** (1989), 168–179.

[117] A. A. UNGAR, *Weakly associative groups*. Resultate Math. **17** (1990), 149–168.

[118] A. A. UNGAR, *The holomorphic automorphism group of the complex disk*. Aequationes Math. **47** (1994), 240–254.

[119] A. A. UNGAR, *Beyond the Einstein Addition Law and its Gyroscopic Thomas Precession: The Theory of Gyrogroups and Gyrovector Spaces*. Kluwer Academic Publishers, Doldrecht-Boston-London 2001.

[120] B. L. VAN DER WAERDEN, *Algebra* I, 5th ed. Springer-Verlag, Berlin-Heidelberg-New York 1960.

[121] H. WÄHLING, *Theorie der Fastkörper*. Thales Verlag, Essen 1987.

[122] W. C. WATERHOUSE, *Self-adjoint operators and formally real fields*. Duke Math. J. **43** (1976), 237–243.

[123] E. L. WILSON, *A class of loops with the isotopy-isomorphy property*. Canad. J. Math. **18** (1966), 589–592.

[124] H. ZASSENHAUS, *Kennzeichnung endlicher linearer Gruppen als Permutationsgruppen*. Abh. Math. Sem. Univ. Hamburg **11** (1936), 17–40.

[125] E. ZIZIOLI, *Fibered incidence loops and kinematic loops*. J. Geom. **30** (1987), 144–156.

[126] M. ZORN, *Theorie der alternativen Ringe*. Abh. Math. Sem. Univ. Hamburg **8** (1930), 123–147.

Index

Mathematical symbols which do not fit naturally into an alphabetical order, are placed at the beginning of the index.

Lecture Notes in Mathematics

For information about Vols. 1–1588
please contact your bookseller or Springer-Verlag

Vol. 1635: E. Hebey, Sobolev Spaces on Riemannian Manifolds. X, 116 pages. 1996.

Vol. 1636: M. A. Marshall, Spaces of Orderings and Abstract Real Spectra. VI, 190 pages. 1996.

Vol. 1637: B. Hunt, The Geometry of some special Arithmetic Quotients. XIII, 332 pages. 1996.

Vol. 1638: P. Vanhaecke, Integrable Systems in the realm of Algebraic Geometry. VIII, 218 pages. 1996.

Vol. 1639: K. Dekimpe, Almost-Bieberbach Groups: Affine and Polynomial Structures. X, 259 pages. 1996.

Vol. 1640: G. Boillat, C. M. Dafermos, P. D. Lax, T. P. Liu, Recent Mathematical Methods in Nonlinear Wave Propagation. Montecatini Terme, 1994. Editor: T. Ruggeri. VII, 142 pages. 1996.

Vol. 1641: P. Abramenko, Twin Buildings and Applications to S-Arithmetic Groups. IX, 123 pages. 1996.

Vol. 1642: M. Puschnigg, Asymptotic Cyclic Cohomology. XXII, 138 pages. 1996.

Vol. 1643: J. Richter-Gebert, Realization Spaces of Polytopes. XI, 187 pages. 1996.

Vol. 1644: A. Adler, S. Ramanan, Moduli of Abelian Varieties. VI, 196 pages. 1996.

Vol. 1645: H. W. Broer, G. B. Huitema, M. B. Sevryuk, Quasi-Periodic Motions in Families of Dynamical Systems. XI, 195 pages. 1996.

Vol. 1646: J.-P. Demailly, T. Peternell, G. Tian, A. N. Tyurin, Transcendental Methods in Algebraic Geometry. Cetraro, 1994. Editors: F. Catanese, C. Ciliberto. VII, 257 pages. 1996.

Vol. 1647: D. Dias, P. Le Barz, Configuration Spaces over Hilbert Schemes and Applications. VII. 143 pages. 1996.

Vol. 1648: R. Dobrushin, P. Groeneboom, M. Ledoux, Lectures on Probability Theory and Statistics. Editor: P. Bernard, 300 pages. 1996.

Vol. 1649: S. Kumar, G. Laumon, U. Stuhler, Vector Bundles on Curves – New Directions. Cetraro, 1995. Editor: M. S. Narasimhan. VII, 193 pages. 1997.

Vol. 1650: J. Wildeshaus, Realizations of Polylogarithms. XI, 343 pages. 1997.

Vol. 1651: M. Drmota, R. F. Tichy, Sequences, Discrepancies and Applications. XIII, 503 pages. 1997.

Vol. 1652: S. Todorcevic, Topics in Topology. VIII, 153 pages. 1997.

Vol. 1653: R. Benedetti, C. Petronio, Branched Standard Spines of 3-manifolds. VIII, 132 pages. 1997.

Vol. 1654: R. W. Ghrist, P. J. Holmes, M. C. Sullivan, Knots and Links in Three-Dimensional Flows. X, 208 pages. 1997.

Vol. 1655: J. Azéma, M. Emery, M. Yor (Eds.), Séminaire de Probabilités XXXI. VIII, 329 pages. 1997.

Vol. 1656: B. Biais, T. Björk, J. Cvitanic, N. El Karoui, E. Jouini, J. C. Rochet, Financial Mathematics. Bressanone, 1996. Editor: W. J. Runggaldier. VII, 316 pages. 1997.

Vol. 1657: H. Reimann, The semi-simple zeta function of quaternionic Shimura varieties. IX, 143 pages. 1997.

Vol. 1658: A. Pumarino, J. A. Rodríguez, Coexistence and Persistence of Strange Attractors. VIII, 195 pages. 1997.

Vol. 1659: V, Kozlov, V. Maz'ya, Theory of a Higher-Order Sturm-Liouville Equation. XI, 140 pages. 1997.

Vol. 1660: M. Bardi, M. G. Crandall, L. C. Evans, H. M. Soner, P. E. Souganidis, Viscosity Solutions and Applications. Montecatini Terme, 1995. Editors: I. Capuzzo Dolcetta, P. L. Lions. IX, 259 pages. 1997.

Vol. 1661: A. Tralle, J. Oprea, Symplectic Manifolds with no Kähler Structure. VIII, 207 pages. 1997.

Vol. 1662: J. W. Rutter, Spaces of Homotopy Self-Equivalences – A Survey. IX, 170 pages. 1997.

Vol. 1663: Y. E. Karpeshina; Perturbation Theory for the Schrödinger Operator with a Periodic Potential. VII, 352 pages. 1997.

Vol. 1664: M. Väth, Ideal Spaces. V, 146 pages. 1997.

Vol. 1665: E. Giné, G. R. Grimmett, L. Saloff-Coste, Lectures on Probability Theory and Statistics 1996. Editor: P. Bernard. X, 424 pages, 1997.

Vol. 1666: M. van der Put, M. F. Singer, Galois Theory of Difference Equations. VII, 179 pages. 1997.

Vol. 1667: J. M. F. Castillo, M. González, Three-space Problems in Banach Space Theory. XII, 267 pages. 1997.

Vol. 1668: D. B. Dix, Large-Time Behavior of Solutions of Linear Dispersive Equations. XIV, 203 pages. 1997.

Vol. 1669: U. Kaiser, Link Theory in Manifolds. XIV, 167 pages. 1997.

Vol. 1670: J. W. Neuberger, Sobolev Gradients and Differential Equations. VIII, 150 pages. 1997.

Vol. 1671: S. Bouc, Green Functors and G-sets. VII, 342 pages. 1997.

Vol. 1672: S. Mandal, Projective Modules and Complete Intersections. VIII, 114 pages. 1997.

Vol. 1673: F. D. Grosshans, Algebraic Homogeneous Spaces and Invariant Theory. VI, 148 pages. 1997.

Vol. 1674: G. Klaas, C. R. Leedham-Green, W. Plesken, Linear Pro-p-Groups of Finite Width. VIII, 115 pages. 1997.

Vol. 1675: J. E. Yukich, Probability Theory of Classical Euclidean Optimization Problems. X, 152 pages. 1998.

Vol. 1676: P. Cembranos, J. Mendoza, Banach Spaces of Vector-Valued Functions. VIII, 118 pages. 1997.

Vol. 1677: N. Proskurin, Cubic Metaplectic Forms and Theta Functions. VIII, 196 pages. 1998.

Vol. 1678: O. Krupková, The Geometry of Ordinary Variational Equations. X, 251 pages. 1997.

Vol. 1679: K.-G. Grosse-Erdmann, The Blocking Technique. Weighted Mean Operators and Hardy's Inequality. IX, 114 pages. 1998.

Vol. 1680: K.-Z. Li, F. Oort, Moduli of Supersingular Abelian Varieties. V, 116 pages. 1998.

Vol. 1681: G. J. Wirsching, The Dynamical System Generated by the 3n+1 Function. VII, 158 pages. 1998.

Vol. 1682: H.-D. Alber, Materials with Memory. X, 166 pages. 1998.

Vol. 1683: A. Pomp, The Boundary-Domain Integral Method for Elliptic Systems. XVI, 163 pages. 1998.

Vol. 1684: C. A. Berenstein, P. F. Ebenfelt, S. G. Gindikin, S. Helgason, A. E. Tumanov, Integral Geometry, Radon Transforms and Complex Analysis. Firenze, 1996. Editors: E. Casadio Tarabusi, M. A. Picardello, G. Zampieri. VII, 160 pages. 1998.

Vol. 1685: S. König, A. Zimmermann, Derived Equivalences for Group Rings. X, 146 pages. 1998.

Vol. 1686: J. Azéma, M. Émery, M. Ledoux, M. Yor (Eds.), Séminaire de Probabilités XXXII. VI, 440 pages. 1998.

Vol. 1687: F. Bornemann, Homogenization in Time of Singularly Perturbed Mechanical Systems. XII, 156 pages. 1998.

Vol. 1688: S. Assing, W. Schmidt, Continuous Strong Markov Processes in Dimension One. XII, 137 page. 1998.

Vol. 1689: W. Fulton, P. Pragacz, Schubert Varieties and Degeneracy Loci. XI, 148 pages. 1998.

Vol. 1690: M. T. Barlow, D. Nualart, Lectures on Probability Theory and Statistics. Editor: P. Bernard. VIII, 237 pages. 1998.

Vol. 1691: R. Bezrukavnikov, M. Finkelberg, V. Schechtman, Factorizable Sheaves and Quantum Groups. X, 282 pages. 1998.

Vol. 1692: T. M. W. Eyre, Quantum Stochastic Calculus and Representations of Lie Superalgebras. IX, 138 pages. 1998.

Vol. 1694: A. Braides, Approximation of Free-Discontinuity Problems. XI, 149 pages. 1998.

Vol. 1695: D. J. Hartfiel, Markov Set-Chains. VIII, 131 pages. 1998.

Vol. 1696: E. Bouscaren (Ed.): Model Theory and Algebraic Geometry. XV, 211 pages. 1998.

Vol. 1697: B. Cockburn, C. Johnson, C.-W. Shu, E. Tadmor, Advanced Numerical Approximation of Nonlinear Hyperbolic Equations. Cetraro, Italy, 1997. Editor: A. Quarteroni. VII, 390 pages. 1998.

Vol. 1698: M. Bhattacharjee, D. Macpherson, R. G. Möller, P. Neumann, Notes on Infinite Permutation Groups. XI, 202 pages. 1998.

Vol. 1699: A. Inoue, Tomita-Takesaki Theory in Algebras of Unbounded Operators. VIII, 241 pages. 1998.

Vol. 1700: W. A. Woyczy´ski, Burgers-KPZ Turbulence, XI, 318 pages. 1998.

Vol. 1701: Ti-Jun Xiao, J. Liang, The Cauchy Problem of Higher Order Abstract Differential Equations, XII, 302 pages. 1998.

Vol. 1702: J. Ma, J. Yong, Forward-Backward Stochastic Differential Equations and Their Applications. XIII, 270 pages. 1999.

Vol. 1703: R. M. Dudley, R. Norvaiša, Differentiability of Six Operators on Nonsmooth Functions and p-Variation. VIII, 272 pages. 1999.

Vol. 1704: H. Tamanoi, Elliptic Genera and Vertex Operator Super-Algebras. VI, 390 pages. 1999.

Vol. 1705: I. Nikolaev, E. Zhuzhoma, Flows in 2-dimensional Manifolds. XIX, 294 pages. 1999.

Vol. 1706: S. Yu. Pilyugin, Shadowing in Dynamical Systems. XVII, 271 pages. 1999.

Vol. 1707: R. Pytlak, Numerical Methods for Optimal Control Problems with State Constraints. XV, 215 pages. 1999.

Vol. 1708: K. Zuo, Representations of Fundamental Groups of Algebraic Varieties. VII, 139 pages. 1999.

Vol. 1709: J. Azéma, M. Émery, M. Ledoux, M. Yor (Eds), Séminaire de Probabilités XXXIII. VIII, 418 pages. 1999.

Vol. 1710: M. Koecher, The Minnesota Notes on Jordan Algebras and Their Applications. IX, 173 pages. 1999.

Vol. 1711: W. Ricker, Operator Algebras Generated by Commuting Projections: A Vector Measure Approach. XVII, 159 pages. 1999.

Vol. 1712: N. Schwartz, J. J. Madden, Semi-algebraic Function Rings and Reflectors of Partially Ordered Rings. XI, 279 pages. 1999.

Vol. 1713: F. Bethuel, G. Huisken, S. Müller, K. Steffen, Calculus of Variations and Geometric Evolution Problems. Cetraro, 1996. Editors: S. Hildebrandt, M. Struwe. VII, 293 pages. 1999.

Vol. 1714: O. Diekmann, R. Durrett, K. P. Hadeler, P. K. Maini, H. L. Smith, Mathematics Inspired by Biology. Martina Franca, 1997. Editors: V. Capasso, O. Diekmann. VII, 268 pages. 1999.

Vol. 1715: N. V. Krylov, M. Röckner, J. Zabczyk, Stochastic PDE's and Kolmogorov Equations in Infinite Dimensions. Cetraro, 1998. Editor: G. Da Prato. VIII, 239 pages. 1999.

Vol. 1716: J. Coates, R. Greenberg, K. A. Ribet, K. Rubin, Arithmetic Theory of Elliptic Curves. Cetraro, 1997. Editor: C. Viola. VIII, 260 pages. 1999.

Vol. 1717: J. Bertoin, F. Martinelli, Y. Peres, Lectures on Probability Theory and Statistics. Saint-Flour, 1997. Editor: P. Bernard. IX, 291 pages. 1999.

Vol. 1718: A. Eberle, Uniqueness and Non-Uniqueness of Semigroups Generated by Singular Diffusion Operators. VIII, 262 pages. 1999.

Vol. 1719: K. R. Meyer, Periodic Solutions of the N-Body Problem. IX, 144 pages. 1999.

Vol. 1720: D. Elworthy, Y. Le Jan, X-M. Li, On the Geometry of Diffusion Operators and Stochastic Flows. IV, 118 pages. 1999.

Vol. 1721: A. Iarrobino, V. Kanev, Power Sums, Gorenstein Algebras, and Determinantal Loci. XXVII, 345 pages. 1999.

Vol. 1722: R. McCutcheon, Elemental Methods in Ergodic Ramsey Theory. VI, 160 pages. 1999.

Vol. 1723: J. P. Croisille, C. Lebeau, Diffraction by an Immersed Elastic Wedge. VI, 134 pages. 1999.

Vol. 1724: V. N. Kolokoltsov, Semiclassical Analysis for Diffusions and Stochastic Processes. VIII, 347 pages. 2000.

Vol. 1725: D. A. Wolf-Gladrow, Lattice-Gas Cellular Automata and Lattice Boltzmann Models. IX, 308 pages. 2000.

Vol. 1726: V. Marić, Regular Variation and Differential Equations. X, 127 pages. 2000.

Vol. 1727: P. Kravanja M. Van Barel, Computing the Zeros of Analytic Functions. VII, 111 pages. 2000.

Vol. 1728: K. Gatermann Computer Algebra Methods for Equivariant Dynamical Systems. XV, 153 pages. 2000.

Vol. 1729: J. Azéma, M. Émery, M. Ledoux, M. Yor, Séminaire de Probabilités XXXIV. VI, 431 pages. 2000.

Vol. 1730: S. Graf, H. Luschgy, Foundations of Quantization for Probability Distributions. X, 230 pages. 2000.

Vol. 1731: T. Hsu, Quilts: Central Extensions, Braid Actions, and Finite Groups. XII, 185 pages. 2000.

Vol. 1732: K. Keller, Invariant Factors, Julia Equivalences and the (Abstract) Mandelbrot Set. X, 206 pages. 2000.

Vol. 1733: K. Ritter, Average-Case Analysis of Numerical Problems. IX, 254 pages. 2000.

Vol. 1734: M. Espedal, A. Fasano, A. Mikelić, Filtration in Porous Media and Industrial Applications. Cetraro 1998. Editor: A. Fasano. 2000.

Vol. 1735: D. Yafaev, Scattering Theory: Some Old and New Problems. XVI, 169 pages. 2000.

Vol. 1736: B. O. Turesson, Nonlinear Potential Theory and Weighted Sobolev Spaces. XIV, 173 pages. 2000.

Vol. 1737: S. Wakabayashi, Classical Microlocal Analysis in the Space of Hyperfunctions. VIII, 367 pages. 2000.

Vol. 1738: M. Émery, A. Nemirovski, D. Voiculescu, Lectures on Probability Theory and Statistics. XI, 356 pages. 2000.

Vol. 1739: R. Burkard, P. Deuflhard, A. Jameson, J.-L. Lions, G. Strang, Computational Mathematics Driven by Industrial Problems. Martina Franca, 1999. Editors: V. Capasso, H. Engl, J. Periaux. VII, 418 pages. 2000.

Recent Reprints and New Editions